Report 142   1994

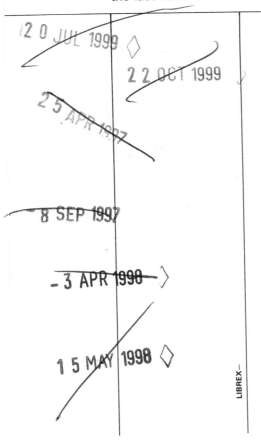

CIRIA   CONSTRUCTION INDUSTRY RESEARCH AND INFORMATION ASSOCIATION
         6 Storey's Gate, Westminster, London SW1P 3AU
         Tel 071-222 8891   Fax 071-222 1708

Summary

It has long been recognised that discharges from surface water drains can be damaging to the environment because urban runoff is likely to contain dissolved and suspended matter washed from roofs, streets, pavements and other surfaces. Runoff from major roads and industrial catchments might also be contaminated with toxic material. More attention is being paid to these discharges as other discharges from sewage treatment works and combined sewer overflows are reduced. There is already considerable attention on discharges to groundwater in the National River Authority's groundwater protection policy.

Recent CIRIA projects have looked at many aspects relevant to highway drainage, and this report should be considered alongside those. They include *Scope for control of urban runoff*, which has assessed runoff control options and *Infiltration drainage - manual of good practice for the design, construction and maintenance of infiltration drainage systems for stormwater runoff control and disposal* (RP448), which gives guidance on the design and maintenance of infiltration systems.

This report reviews the pollutants that are likely to be present in highway drainage discharges, and assesses the impact that these can have on the receiving water. It also reviews the legal framework within which the discharges are made, and the controls that can be imposed on these discharges. It also reviews current drainage practice and the operation of different types of drainage structure in controlling pollution.

The report then presents guidelines for the planning and design of highway drainage systems so as to avoid pollution of the receiving waters. These guidelines are intended for immediate practical use by highway drainage designers, and water quality regulators.

The report concludes with recommendations for future work to lead to a better understanding of pollution from highway drainage, and for improved practice to reduce the environmental impact of the drainage discharges.

Control of pollution from highway drainage discharges
Construction Industry Research and Information Association Research Project 473, 1994

Keywords
Highway runoff, urban drainage, water quality, pollution

Reader interest
Highway and road drainage engineers, highway and road maintenance authorities, water quality planners, pollution control authorities, environmental scientists and engineers.

All rights reserved. No part of this publication may be reproduced or transmitted in any form or by any means, including photocopying and recording, without the written permission of the copyright holder, application for which should be addressed to the publisher. Such written permission must also be obtained before any part of this publication is stored in a retrieval system of any nature.

© CIRIA 1994

ISBN 086017 415 8

CLASSIFICATION	
AVAILABILITY	Unrestricted
CONTENT	Practical guidance
STATUS	Committee guided
USER	Drainage engineers and water quality scientists

Published by CIRIA, 6 Storey's Gate, Westminster, London SW1P 3AU

Foreword

This report was produced as a result of CIRIA Research Project 473, *Control of pollution from new and existing highway drainage systems*, carried out on behalf of CIRIA by Thorburn Colquhoun. This document also constitutes NRA R & D Report 16.

The main body of the report was written by Messrs Martin Luker and Keith Montague of Thorburn Colquhoun. Contributions to the main body of the text were received from:

      Mr Roderic Cameron of Water Management Consultants
      Mr Mike Price of Reading University
      Mr Nick Walton of Portsmouth University
      Mr Philip Burton of Brian Colquhoun and Partners
      Drs Nigel Graham, David Butler and Christopher Sollars of Imperial College.

Section 3 of the report was written by Professor William Howarth, Cripps Harries Hall/SAUR, Chair of Environmental Law at the University of Kent, Canterbury. Contributions to Section 3 were received from:

      Mr Michael R Flegg and colleagues of Cripps Harries Hall solicitors
      Mr Donald McGillivray of the University of Kent.

Appendix A of the report was written by Mr Martin Osborne of CIRIA.

CIRIA's Research Managers were Dr Judy Payne and Ms Siân John. Following CIRIA's usual practice, the work was guided by a Steering Group.

Steering Group

Chairman

Mr Mike LeGouais	Scott Wilson Kirkpatrick & Partners

Members

Mr Peter Allen	Department of Transport
Mr Ron Boots	Department of Transport
Mr Doug Colwill	Transport Research Laboratory
Professor Bryan Ellis	Middlesex University
Mr Paul Freckleton	Yorkshire Water Services
Dr John Gardiner	National Rivers Authority
Mr Paul Johnson	Transport Research Laboratory
Mr Gerard Morris	National Rivers Authority
Mr Grahame Newman	British Waterways Board
Mr Martin Squibbs	Tayside Regional Council
Mr Bob Wilkins	Hertfordshire County Council, representing the Local Authority Associations
Mr John Wood	WRc

Funders

CIRIA is grateful to the following organisations who provided financial support for the project:

British Waterways	Central Regional Council
Department of Transport	Grampian Regional Council
Lothian Regional Council	National Rivers Authority
Tayside Regional Council	

CIRIA and the authors also wish to thank everyone else who contributed to the contents of the report and all the organisations and individuals who took part in the questionnaire survey. Also the funders of WRc project U-1018 *Surface water outfalls quality and environmental management*, who agreed to a free exchange of information between the two projects and are listed below:

Yorkshire Water Limited   Strathclyde Regional Council (Sewerage)
DoE Northern Ireland

The permission of HR Wallingford to reproduce figures from "The Wallingford Procedure - Volume 1" as Figures 5.2 and 5.4 is gratefully acknowledged.

Contents

List of tables — 10

List of figures — 10

Abbreviations — 12

1 INTRODUCTION — 15
  1.1 How to use this report — 15
  1.2 Background — 15
  1.3 Objectives — 16
  1.4 Scope and limitations — 16
  1.5 Methodology — 17

2 POLLUTION FROM HIGHWAY DRAINAGE SYSTEMS — 18
  2.1 Classification of pollutants — 18
    2.1.1 Sediments — 18
    2.1.2 Hydrocarbons — 18
    2.1.3 Metals — 19
    2.1.4 Salts and nutrients — 20
    2.1.5 Microbial — 20
    2.1.6 Others — 20
  2.2 Sources of pollution — 21
  2.3 Pollution from routine discharges — 22
    2.3.1 Traffic — 22
    2.3.2 Maintenance — 25
    2.3.3 Other sources of pollution — 27
    2.3.4 Total loadings — 28
    2.3.5 Quality of runoff — 30
  2.4 Accidental spillages — 32
    2.4.1 Types of spillage — 32
    2.4.2 Number of incidents — 34
  2.5 Scale of pollution problems — 35
    2.5.1 Pollution of surface water — 35
    2.5.2 Pollution of groundwater — 36
    2.5.3 Aquifer recharge — 38

3 STANDARDS, LEGISLATION AND PRACTICE — 41
  3.1 Introduction and overview — 41
    3.1.1 Introduction — 41
    3.1.2 Overview — 41
  3.2 Legal powers and duties — 42
    3.2.1 The National Rivers Authority in England and Wales — 42
    3.2.2 River Purification Authorities in Scotland — 46
    3.2.3 Highway responsibilities in England and Wales — 48
    3.2.4 Roads responsibilities in Scotland — 50
    3.2.5 The transport of polluting substances by road — 54
  3.3 The European Community and national water quality law — 55
    3.3.1 European Community law in England and Wales — 55
    3.3.2 National Water Quality Objectives in England and Wales — 57
    3.3.3 European Community law and Water Quality Objectives in Scotland — 59
  3.4 Water pollution law — 59
    3.4.1 Water pollution offences in England and Wales — 59
    3.4.2 Water pollution offences in Scotland — 62
    3.4.3 Defences in England and Wales — 62
    3.4.4 Defences and authorisations in Scotland — 64
    3.4.5 Discharge consents in England and Wales — 65
    3.4.6 Discharge consents in Scotland — 66

3.5 Water resource protection	66
3.5.1 Water resources and the National Rivers Authority	66
3.5.2 Water resource protection in Scotland	68
3.5.3 The Civil Law	68
4 CURRENT PRACTICE AND CONTROL OPTIONS	**71**
4.1 The function of highway drainage	71
4.2 General design considerations	71
4.2.1 Runoff	71
4.2.2 Design criteria and techniques	71
4.2.3 Pollution	72
4.3 Highway drainage methods and drainage system components	73
4.3.1 Highway drainage functions	73
4.3.2 Kerbs and gully pots	76
4.3.3 Filter drains	77
4.3.4 Road-edge surface water channels	79
4.3.5 Porous surfacing	81
4.3.6 Precast channel or slot drains	81
4.3.7 Informal verge systems	82
4.3.8 Fin drains	83
4.3.9 Infiltration pavements	84
4.3.10 Catchpits, grit traps and manholes	85
4.3.11 Oil separators	86
4.3.12 Swales	87
4.3.13 Infiltration basins	88
4.3.14 Soakaways and infiltration trenches	89
4.3.15 Storage ponds and detention tanks	91
4.3.16 Sedimentation tanks	94
4.3.17 Lagoons	95
4.3.18 Wetlands	97
4.4 Control of problem pollutants	98
4.5 Pollution management	99
4.5.1 Pollution control through cleaning operations	99
4.5.2 Pollution control as part of winter maintenance operations	101
4.5.3 Pollution control during other maintenance operations	102
4.5.4 Pollution control at source	102
5 GUIDANCE	**104**
5.1 Introduction	104
5.2 Characteristics of highway runoff	104
5.2.1 Sources of pollutants	104
5.2.2 Pollutants of major concern	104
5.2.3 Quantities of pollutants	107
5.2.4 Washoff of pollutants	107
5.3 Water quality standards	108
5.3.1 Framework	108
5.3.2 Groundwater	108
5.3.3 Surface waters	110
5.4 Guide to good practice	112
5.4.1 Groundwaters	112
5.4.2 Surface waters	114

5.5 Appropriate drainage techniques	121
5.5.1 Gully pots	121
5.5.2 Combined surface water and groundwater filter drains	121
5.5.3 Informal verge systems	124
5.5.4 Infiltration basins	125
5.5.5 Swales (grass channels)	125
5.5.6 Soakaways and infiltration trenches	125
5.5.7 Storage ponds and detention tanks	125
5.5.8 Catchpits/grit traps	126
5.5.9 Oil separation	126
5.5.10 Sedimentation tanks	126
5.5.11 Wetlands	126
5.5.12 Lagoons	127
5.5.13 Maintenance implications	127
6 CONCLUSIONS	128
6.1 Pollutants in runoff	128
6.2 Legislation	129
6.3 Design issues	129
6.4 Maintenance	129
7 RECOMMENDATIONS	131
7.1 Procedural issues	131
7.2 Drainage techniques	131
7.3 Maintenance	132
7.4 Design and assessment methods	133
REFERENCES	134
APPENDIX A DEVELOPMENT OF NUMERICAL METHOD FOR SURFACE WATERS	142
A.1 Background to the method	142
A.1.1 Scope	142
A.1.2 Outline of the method	142
A.2 Development of the method	143
A.2.1 Pollutant washoff model	143
A.2.2 Test data sets	144
A.2.3 Initial analysis	144
A.2.4 Sensitivity to timestep	145
A.2.5 Sensitivity to erosion rate	145
A.2.6 Identification of critical storms for Site A	145
A.2.7 Analysis of Site B data	146
A.2.8 Analysis of the Site C data	146
A.3 Development of a general method	146
A.3.1 Rainfall statistics	147
A.3.2 Testing general method against derived results	148
A.3.3 Comparison with water quality standards	148
APPENDIX B HIGHWAY DRAINAGE ASSESSMENT WORKSHEET	150
B.1 Worked example	151

List of Tables

Table 2.1	*Sources and classifications of pollution*	21
Table 2.2	*Tyre abrasion on urban streets*	24
Table 2.3	*Emissions from abrasion of urban streets*	24
Table 2.4	*Emissions due to abrasion of brake linings on urban streets*	24
Table 2.5	*Annual pollutant loadings from highway traffic*	29
Table 2.6	*Reported ranges of pollutant concentrations found in various locations*	31
Table 2.7	*Extracted data from NRA (Thames Region) pollution incident register*	34
Table 4.1	*Functions and water quality attributes of highway drainage methods*	75
Table 4.2	*Trap efficiency of 450mm diameter BS gully pot*	76
Table 4.3	*Mean annual removal efficiencies (%) for a filter drain, sedimentation tank and lagoon at an M1 experimental site*	79
Table 4.4	*Characteristics and removal rates for grass swales*	88
Table 4.5	*Removal efficiencies for French motorway infiltration basins*	89
Table 4.6	*Definition of surface storage pond types*	92
Table 4.7	*Detention basin removal efficiencies*	93
Table 4.8	*Percentage removal efficiencies for dry type oil removal*	96
Table 4.9	*Performance of experimental reed bed systems*	98
Table 4.10	*Efficiency of sweeping*	100
Table 5.1	*Typical pollutant build-up rates*	107
Table 5.2	*Pollutants per outing of road salting*	107
Table 5.3	*Standards for water for human consumption*	110
Table 5.4	*Standards for abstraction of drinking water*	111
Table 5.5	*Standards for fisheries ecosystems*	111
Table 5.6a	*Dissolved pollutant abatement requirements*	119
Table 5.6b	*Aesthetic pollution abatement requirements*	119
Table 5.7	*Ratings table for management measures*	122
Table 5.8	*Applicability of drainage measures to highway situations*	123
Table A.1	*Pollutant concentrations for daily and hourly data for Site A*	145
Table A.2	*Classification of critical storms for Site A*	146
Table A.3	*Classification of storms for Site B*	146
Table A.4	*Classification of storms for Site C*	146
Table A.5	*Comparision with rainfall statistics*	147
Table A.6	*Pollutant concentrations for general method and rainfall series analysis*	148
Table A.7	*Comparision of predicted concentrations with fisheries standards for copper*	149

List of Figures

Figure 2.1	*Average lead in air, lead in petrol and flow of petrol engined vehicles*	23
Figure 2.2	*Induced recharge from river to aquifer*	39
Figure 4.1	*Typical installation of a trapped gully with kerb*	76
Figure 4.2	*Typical installation of a chute type gully with kerb*	77
Figure 4.3	*Cross section of a typical filter drain*	79
Figure 4.4	*Section view of surface water channel outfall*	80
Figure 4.5	*Typical cross section of porous surfacing draining to filter drain*	81
Figure 4.6	*Proposed application of slot drain for drainage of porous surfacing*	82
Figure 4.7	*Fin drain in edge of pavement*	83
Figure 4.8	*Typical cross sections of infiltration pavements*	84
Figure 4.9	*Typical bypass type oil separator*	86
Figure 4.10	*A French filtration basin for treatment of motorway runoff*	89
Figure 4.11	*Section through soakaway*	90
Figure 4.12	*Cross section of infiltration trench*	90
Figure 4.13	*Typical on-stream and off-stream storage ponds*	92
Figure 4.14	*Cross section of an experimental treatment lagoon*	96
Figure 4.15	*Typical arrangement for a reed bed treatment system*	97
Figure 5.1	*Procedure for assessing discharges to groundwaters*	113

Figure 5.2 *Average annual rainfall 1941-1970*   115
Figure 5.3 *Procedure for assessing discharges to surface waters*   116
Figure 5.4 *Depth of rain for assessing pollutant washoff*   118
Figure 5.5 *Typical cross section of an improved filter drain*   124

Abbreviations

GENERAL

AADT	Annual Average Daily Traffic
ACC	Association of County Councils
ADA	Association of District Authorities
AMA	Association of Metropolitan Authorities
BGS	British Geological Survey
BOD	Biochemical Oxygen Demand
BS	British Standard
BSI	British Standards Institute
CFU	Colony Forming Units
CIA	Chemical Industries Association
CIRIA	Construction Industry Research and Information Association
COD	Chemical Oxygen Demand
COSHH	Control of Substances Hazardous to Health Regulations
CSO	Combined Sewer Overflow
DDT	Dichloro-Diphenyl-Trichloroethane
DMRB	Design Manual for Roads and Bridges
DoE	Department of the Environment
DOT	Department Of Transport
EC	European Community
Eh	Negative logarithm of electron activity
EIFAC	European Inland Fisheries Advisory Commission
EQS	Environmental Quality Standards
FHA	Federal Highways Administration (USA)
GAC	Granular Activated Carbon
HA nn/nn	Highways Advice note
HECB	Highway Engineering Computer Branch (DOT)
HGV	Heavy Goods Vehicle
HR	HR Wallingford
HRS	Hydraulics Research Station (now HR Wallingford)
ICI	Imperial Chemical Industries
ICPS	Inductively coupled plasma spectrophotometry
LAA	Local Authorities Association
MTBE	Methyl-Tertiary-Butyl-Ether
NRA	National Rivers Authority
PAH	Polynuclear Aromatic Hydrocarbons
PCB	PolyChlorinated Biphenyl
RIDDOR	Reporting Injuries Diseases and Dangerous OccuRrences
RP nnn	Research Project (CIRIA)
SS	Suspended Solids
TN	Total Nitrogen
TP	Total Phosphorus
TPb	Total Lead
TRL	Transport Research Laboratory
TRRL	Transport and Road Research Laboratory (now TRL)
TS	Total Solids
TSS	Total Suspended Solids
TZn	Total Zinc
UPM	Urban Pollution Management programme
XRF	X-Ray Fluorescence (spectrophotometry)

LEGAL ABBREVIATIONS

SI	Statutory Instrument
ss.	Sections
Art.	Article
Reg.	Regulation

The system of abbreviation used in Section 3 is the formal system of referencing for all legal documentation as applied to the United Kingdom.

Example {s.100 (4)(iii) HA 1980}

This is a reference to section 100, sub-section 4, sub-sub-section iii of the Highways Act 1980

Civil Law references are given in full detail to allow the reader to access the relevant reference.

ENACTMENTS FREQUENTLY REFERRED TO

National Legislation

Control of Pollution Act 1974	**COPA1974**
Health and Safety at Work etc. Act 1974	**HSWA1974**
Highways Act 1980	**HA1980**
Rivers (Prevention of Pollution) (Scotland) Act 1951	**R(PP)(S)A1951**
Roads (Scotland) Act 1984	**R(S)A1984**
Town and Country Planning Act 1990	**TCPA1990**
Town and Country General Development Order	**GDO1988**
Water Industry Act 1991	**WIA1991**
Water Resources Act 1991	**WRA1991**
Water (Scotland) Act 1980	**W(S)A1980**

European Directives

Dangerous Substances Directive {76/464/EEC}	**DSD1976**
Groundwater Directive {80/68/EEC}	**GD1980**

TERMINOLOGY

The report generally uses terminology applicable to the organisation and law in England and Wales. For use in Scotland, the words "highway" and "street" have the same meaning as "road".

1 Introduction

1.1 HOW TO USE THIS REPORT

This report is intended for use by practising highway and development engineers, responsible for the design of new highway drainage systems. It is not specifically intended for modifications to existing systems but, where techniques can be used to improve an existing situation, this is noted in the text. It is also intended to act as a guide to the staff of regulatory authorities who need to consider the impact of highway discharges.

In **Section 2**, the forms, composition and sources of pollution are discussed.

In **Section 3**, existing legislation is reviewed and the current duties and responsibilities of highway and pollution control authorities are set out, including the effects of EC Directives.

Section 4 deals with the capability of existing drainage technology to counter the threat of pollution.

Section 5 summarises Sections 2 and 4 and presents guidance for the design of highway drainage systems.

Section 6 presents the conclusions of this stage of the study.

Section 7 makes recommendations including future research needs.

The report is intended to be understood by and of use to the non-specialist. Designers of highway drainage systems may need to refer only to the guidance sections of this report:

- Section 1   Introduction
- Section 5   Guidance
- Section 6   Conclusions
- Section 7   Recommendations.

1.2 BACKGROUND

Current drainage practice in Britain is to build separate foul and surface water sewerage systems. These have been in use in Britain since 1950 when they were introduced in preference to combined and partially separate systems. The principal argument in favour of separate systems, and that which justified the additional costs, was that environmentally harmful sewage discharges to watercourses would be substantially reduced.

As the problems of combined sewer overflows are being addressed and the consequent pollution greatly reduced, attention is now being turned to discharges from surface water drains that may themselves be damaging to the environment. This is because surface water runoff contains dissolved and suspended matter washed from roofs, streets, pavements and other surfaces. It has been suggested that runoff from major roads may be a substantial contributor to contamination, because by far the majority of separate surface water discharges are believed to be either highway drains or drains that carry considerable quantities of highway runoff.

CIRIA Research Project 404 *Scope for control of urban runoff* (published in 1992) assessed the options for the control of runoff quantity both at source and in the drainage system. This is being followed up by RP 448 *Manual on infiltration methods for stormwater source control*, which will produce guidance on the design and maintenance of infiltration systems. Project RP 416 *Cleaning sediment from sewerage and drainage systems and related above-ground areas* is examining the effects of street cleaning and gully emptying on sediments, which contribute to pollution, entering drains. Similarly, guidance is needed by highway authorities, environmental regulators and developers on: the impact of highway drainage discharges; the need for measures to reduce resultant pollution; the types of drainage system that can be constructed and their application in the construction of new or refurbished highways.

The roads of the British Isles in all classifications total some 360 000km and in 1991 there were approximately 24.5 million vehicles using those roads. The number of vehicles is expected to continue to grow and Department of Transport (DOT) statistics predict that by the year 2010 there may be as many as 40 million.

The volume of detritus produced from all sources by traffic movements, plus all the attendant leakages of fuel, oil and other hydrocarbon products, results in measurable quantities of contaminants being washed into the aquatic environment. If there is insufficient natural dilution, consideration should be given to methods that reduce contaminant concentrations to acceptable levels.

1.3 OBJECTIVES

The overall objective of this research project was to determine the impact of highway runoff and to produce practical guidance on measures that can be taken to control this pollution in Great Britain. The research has been structured in two stages. This report concludes Stage 1 of the project, which had the following specific objectives:

- to determine and document the nature and scale of problems caused by discharges of highway runoff
- to describe the legal, administrative and regulatory framework for the management and control of highway drainage in England and Wales and road drainage in Scotland (including European legislation and standards)
- to summarise existing technical knowledge and describe current practices for the control of pollution from highway drains
- to identify possible solutions to the problems caused by discharges from highway drains and to develop guidance
- to identify and prioritise research and development needs for the control of highway runoff discharges, including consideration of pollution arising from normal road use, spillages and maintenance activities
- to produce a detailed proposal for Stage 2 of the project.

1.4 SCOPE AND LIMITATIONS

The National Rivers Authority (NRA) in England and Wales and the Regional Councils in Scotland are working together with the DOT and other highway authorities to look at the impact of highways on the aquatic environment. This project is one of several being carried out within this joint research and development liaison framework.

The scope of the report is as set out in the objectives with attention focused on highway drainage systems; public sewerage is not considered. The report concerns runoff from major urban and rural roads but not specifically from estate roads and parking areas, and does not cover the construction phase. The potential impact of runoff discharges on surface waters, both natural and man-made, and on groundwater, is assessed.

The guidance given is based on a distillation of current best practice and the development of an outline design method. Proposals are included for further testing and development of these guidelines.

1.5 METHODOLOGY

The work was carried out by a literature review and a follow-up questionnaire survey, supplemented by discussions with practitioners. The authorities contacted included Water plcs, NRA regional offices, the Scottish Office, Scottish Regional Councils, DOT regional offices, County Councils, Metropolitan Boroughs, District Councils, Consulting Engineers and Water Companies. Three hundred and twenty four questionnaires were distributed and 187 returns were received. The response rate of 58% was encouraging and taken to indicate a positive interest in the project amongst the design and regulatory parties.

2 Pollution from highway drainage systems

2.1 CLASSIFICATION OF POLLUTANTS

Although the public perceives pollution as being a single definable quantity it is in fact a complex matrix of interrelated substances. For example, metals come from a number of sources and exist in highway discharges in several different forms i.e. soluble, particulate solids and salts. If each of these forms was to be considered in isolation then the involvement of inorganic sediment as a transport medium could be overlooked. To clarify this complex issue, a classification system has been developed for use in this report. The system is not intended to be either rigorous or exhaustive, but has the merit that it can be directly related to the techniques for treatment of highway discharges.

The system divides pollutants in highway drainage discharges into the following six categories:

- sediments
- hydrocarbons
- metals
- salts and nutrients
- microbial
- others.

2.1.1 Sediments

Sediment is most simply defined as material that settles to the bottom of a liquid, i.e. material of a higher specific gravity than water. Particle size is important. Coarse particles will not be carried in suspension even in fast flows; they are transported by the process of saltation (rolling and bouncing along the bed). Fine particles may be so fine that they will only settle out of suspension after long periods of low-energy flow, or as a result of natural coagulation, when gravitational forces overcome the forces of molecular repulsion.

The chemical nature of sediments varies considerably and is of equal importance to particle size. Even inert sediments can have a significant effect on surface waters as they can blanket the bed and affect fish. Extence (1978) and others have found that inert sediments are the main cause of lower biotic indices in the receiving watercourses.

The greatest potential threat from sediment, in terms of pollutant load, is the manner in which it can act as a transport medium or a coagulant for materials such as hydrocarbons, heavy metals, salts and nutrients. Sediment also includes organic detritus, which has an impact due to the oxygen demand that it exerts when discharged into receiving waters. Research has indicated that the fraction of sediment smaller than 63 μm is the most significant for pollution (Sartor & Boyd, 1972; Laxen & Harrison, 1977; Ellis, 1979). Although this may be only 6% of the total mass of sediment, it can constitute up to 50% of the pollution load of associated metals, hydrocarbons, COD, nutrients and herbicides (Collins and Ridgeway, 1980).

Sediments are not usually a problem in groundwater as they are filtered out before the discharge reaches the water table. However, the filtered sediments may still cause a problem by continuing to leach pollutants into the water table.

2.1.2 Hydrocarbons

In this report the term hydrocarbons is used to mean organic compounds containing only carbon and hydrogen, particularly the petrochemical derived group, which includes petrol, fuel oils, lubricating oils and hydraulic fluids. In unmodified form these are liquid, virtually insoluble and lighter than water. Some hydrocarbons, such as bitumen and heavy fuel oil, become heavier than water when affected by naturally occurring bacteria and are then classifiable as sediments. Typically 70-75% of hydrocarbon oils show a strong attachment for the suspended sediments. If most of the suspended solids were removed, the remaining oil/hydrocarbon levels would be around a few mg/l. Polynuclear Aromatic Hydrocarbons (PAHs) have an even higher affinity (Ellis, 1991b). Conversely the new MTBE (methyl-tertiary-butyl-ether) additive to unleaded fuel is significantly more soluble in water than all the other automotive hydrocarbons.

Even low concentrations of hydrocarbons can cause problems in surface waters as they give rise to surface sheens. Hydrocarbons also impart tastes and odours to both surface and groundwaters and can render them unsuitable for water abstraction. Hydrocarbons are degraded by a combination of microbial and oxidative processes, which gradually reduces the impact on the watercourse, although there will be an oxygen demand during their breakdown.

Forrow et al (1993) looked in particular at the PAH fraction of hydrocarbons and found a variety of compounds in the 2-20 $\mu g/l$ range mostly attached to sediments. They found significant reductions in biotic indices downstream of the runoff input but could not assign this specifically to their identified PAH or other hydrocarbon pollutants. However, in a recent extension of this work, Boxall et al (1993), showed that aromatics extracted from sediments were toxic to the sensitive freshwater shrimp *Gammarus pulex*, and that the main causes were the PAHs, Pyrene and Fluoranthene, which were found in the tissue of *Gammarus pulex* at levels of 5-8 $\mu g/g$.

2.1.3 Metals

The majority of studies on metals in highway runoff have concentrated on lead, cadmium, copper, zinc and iron. Some studies have also included nickel, chromium and manganese. The more unusual metals, such as titanium, vanadium, cobalt, arsenic, molybdenum, tin, tungsten and antimony, are occasionally mentioned as trace constituents. The unusual metals are more likely to be included in recent studies due to the increasing availability of rapid, multi-element analytical techniques like ICPS and XRF.

Metals can exist in many forms, modified or unmodified. They can be attached to inert sediments, or be contained in immiscible fluids, or occur as particles, soluble salts or insoluble compounds. Chemically they can be organic or inorganic, compounds or complexes, and can usually exist amongst a variety of ionic species depending primarily on the prevailing redox and pH conditions. Research indicates that the metals in highway runoff are predominantly in or associated with the particulate phase. This is a crucial point since the environmental mobility and bio-availability of metals is largely dependent upon their concentration in solution. Conversely, most analytical measurements produce results for total metal concentration, although some studies do specifically state filter size or solvent used to distinguish between various forms of the metals present. A number of studies have sought to quantify the soluble portion with the following typical results:

Lead	1-10% soluble
Copper	20-40% soluble
Zinc	30-50% soluble

Levels of soluble metals will tend to be highest and toxicity thresholds lowest in soft water (low pH) areas such as in the non-calcareous highland areas of Scotland, Wales, north-west and south-west England, whereas the opposite is the case for the hard water areas of central, southern and eastern England. Morrison et al (1988) showed how developing anoxic conditions in gully pot storage could alter the local Eh/pH environment sufficiently to liberate metals bonded to the particulate matter, into solution. These metals would then be flushed out into the outfall with the next storm input to the gully pot.

Cadmium is a very toxic metal that accumulates in the environment. It is present in highway runoff, but its use for all purposes is now restricted, and so the concentrations are reducing.

Lead is also a serious and accumulative poison. Low concentrations of soluble lead may affect tadpoles, frogs (very sensitive) and fish such as minnows, stickleback and trout (0.1-6.4 mg/l may be toxic). The ecological impact of lead is significantly less than might be expected because of its low solubility.

Fish are also particularly sensitive to dissolved copper and zinc at sub mg/l levels.

Iron, although not toxic, can cause discolouration and other physical problems when present at high concentrations. It is also a useful total metals indicator for monitoring purposes. Iron oxides (rust) have a particularly high surface area affinity for a number of trace metals such as lead, cadmium, arsenic, etc. so that the presence of iron oxides can exert a beneficial scavenging effect on other metals in highway runoff.

2.1.4 Salts and nutrients

Salts and nutrients are defined as those generally neutral materials that occur as soluble compounds and have a direct polluting effect upon vegetable matter either by reducing, or extinguishing conditions conducive to propagation or by accelerating growth to the detriment of the balance of the environment.

In North America chloride from highway de-icing has been widely reported as a source of contamination of both groundwater and surface waters (Saleem, 1977; Howard & Beck, 1993) and its use has been banned or restricted in some areas. Chloride is known to be present in high concentrations in runoff from highways in Britain during winter (Hedley & Lockley, 1975; Bellinger et al, 1982; Colwill et al, 1984). However there have been no specific reports of increased chloride levels in British groundwaters.

Undesirable algal growths may occur downstream of discharges containing nutrients and reduce light incidence below the surface. Blue-green algae also produce high levels of toxins, and all algae can cause oxygen demand and release of toxins when they die. Dussart (1984) reports an increase in algae downstream of discharges from the M6 in Cumbria and north Lancashire.

Hvitved-Jacobsen et al (1986) found that 99% of phosphorus and 85-90% of nitrogen nutrient loadings are removed with the sediment, indicating that excess nutrient loadings into receiving watercourses are unlikely to be a problem if the sediments are effectively removed.

2.1.5 Microbial

Microbial activity is mainly associated with the particulate material derived from the decay of organic matter or finely divided solids that harbour bacteria or viruses. Significant microbial populations are transported with wind blown soils.

2.1.6 Others

Other substances do not readily fit into the other classes. Examples of these materials are pesticides and herbicides.

Herbicides and pesticides can be toxic to a variety of aquatic life at very low concentrations. Some of the more toxic varieties e.g. the chlorinated organics, such as DDT and PCBs, are no longer in use but their residues can still be found both in atmospheric dusts and in areas where they have been used in the past. The previously widespread use of triazine herbicides, including simazine and atrazine, is now also being phased out but their residues will persist for some considerable time. However, whatever chemicals are used in the future, no matter how target-specific they become or how tightly controlled their application, they may present an environmental problem to the local ecology of the receiving water.

In surface water courses, pesticides may become concentrated in the food chain by being retained in certain tissues or parts of organisms. Conversely, they can become selectively adsorbed onto certain materials.

Under current regulations (DoE 1989b) based on EC Directive 80/778/EEC, herbicides and pesticides are not permitted in public water supplies in concentrations greater than 0.1 μg/l for any individual pesticide, or 0.5 μg/l in total. These limits are not based on toxicological evidence but on the levels of detectability at the time the directive was drafted. It has been argued that they are unnecessarily strict. Removal of these compounds at treatment works is costly, as it is normally carried out using granular activated carbon (GAC).

2.2 SOURCES OF POLLUTION

Pollutants contained in highway runoff can be generated from a wide variety of sources, which, for convenience, have been divided into four main groups, as follows:

- *traffic* from the operation and passage of motor vehicles including those arising from abrasion, corrosion and attrition of both vehicles and highway surfaces
- *maintenance* from maintenance operations carried out on roads (e.g. de-icing, defoliation)
- *accidental discharges* from accidents and spills
- *others* from miscellaneous other sources, e.g. atmospheric deposition, maintenance of vehicles, illegal disposal, agricultural activities.

Table 2.1 indicates the complexity, range and classification of potential contaminants, which are listed under the four source group headings.

Table 2.1 Sources and classifications of pollution

Sources / Classification	Traffic	Maintenance	Accidents and disposal	Other
Sediment	carbon organic solids rubber litter and plastics grit asbestos	de-icing grit re-surfacing grit rust metal filings plastic organic solids	soil cement sand gravel plastic litter	atmospheric dust organic detritus soil
Metals	lead zinc nickel and chromium iron and rust cadmium copper vanadium	iron copper chromium nickel cadmium arsenic	lead zinc copper chromium iron	iron manganese
Hydrocarbons	petrol and oil PAHs and MTBE hydraulic fluid grease antifreeze olefins	tar & bitumen asphalt PAHs oils grease solvents	petrol solvents oil grease	
Salts and nutrients	nitrates bromide sulphates ammonia	chlorides sulphates phosphates urea bromide cyanide	fertiliser (N&P salts) organo-compounds	acid rain
Microbial	livestock movements excrement	organic solids	bacteria viruses organic solids excrement	bacteria viruses flesh, blood and bone
Others		PCBs, herbicides and pesticides	herbicides and pesticides	

2.3 POLLUTION FROM ROUTINE DISCHARGES

2.3.1 Traffic

The movement of traffic on highways generates polluted runoff as a result of:

- *emissions* from the vehicles themselves, due to the workings of the internal combustion engine, i.e. combustion and partial combustion of hydrocarbon fuels, which include alkyl lead in leaded petrol and olefins or MTBE additives in unleaded petrol
- *abrasion and corrosion* as a result of the movement of vehicles, affecting both the vehicle and the highway surface
- *turbulence* caused by vehicle movement, activating existing materials.

Vehicle emissions

Vehicle emissions include volatile solids and Polynuclear Aromatic Hydrocarbons (PAHs) derived from unburnt fuel, exhaust gases and vapours, lead compounds (from petrol additives) and hydrocarbon losses from the fuel, lubrication and hydraulic systems. Volatile solids (except for the finest fractions that will be lost) will be added to the total suspended solids loading of highway runoff and can also act as carriers for both metals and hydrocarbons. Minor quantities of bromide and nitrate may also be included in exhaust emissions (Colwill et al, 1984; Bellinger et al, 1982). Exhaust gases and vapours will only contribute when they are taken up by rainfall or snow and returned to the road surface as a component of atmospheric deposition.

The emission of lead from vehicle exhausts has been an acknowledged problem for some time (Little & Wiffen, 1977;1978). The tetra-alkyl organic compounds of lead have been routinely added to petrol for more than 50 years to boost the octane rating, thereby increasing engine efficiency by acting as anti-knock agents. After combustion, these compounds are exhausted in fine particulate form, mainly in two distinct size ranges of less than 1 μm and between 5 and 50 μm (Laxen and Harrison, 1977).

The actual rate and form of lead emission is crucially dependent on driving conditions and circumstances (Laxen and Harrison, 1977). High engine speeds and rapid acceleration will increase emission rates, the latter especially because of reactivation of particles deposited in the exhaust system. Engines under stress also emit a higher proportion of larger particles, whereas the finer fractions predominate under cruise conditions. Total lead emissions can double when the engine and/or the weather is cold. Various emission rates have been quoted e.g. 0.029 g/vehicle km while cruising at 110 km/h and up to 0.068 g/vehicle km under acceleration (Hedley and Lockley, 1975). It should be noted, however, that the permitted concentration of lead added to petrol in Britain has declined during the 1980s from 0.45 g/l to 0.15 g/l, while sales of unleaded petrol have risen to 52% of the total (DoE, 1993b) and are expected to continue to increase because of the increasingly strict environmental legislation and the financial inducements of reduced levels of duty payable (see Figure 2.1). Lead emission studies published to date should therefore be used with care.

The replacement additives for alkyl lead in "unleaded" fuel are Olefins and MTBE. The products of combustion of these additives are gaseous rather than particulate and so should not contribute to pollution from highway drainage. However, research should be carried out into the concentration of these materials in highway runoff.

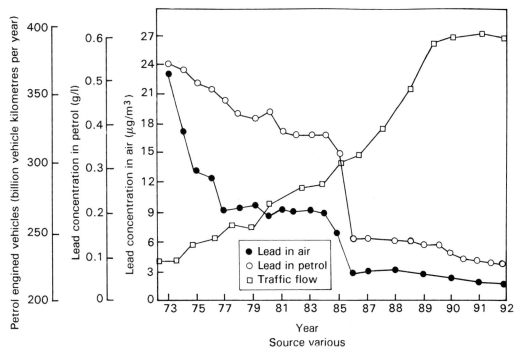

Figure 2.1 *Average lead in air, lead in petrol and flow of petrol engined vehicles*

On main roads, leaks from vehicle lubrication and hydraulic systems provide a steady source of fluid hydrocarbons. Lubricating oils exert an oxygen demand and may also contain organic phosphates and metals, the latter derived from the engine parts (Dussart, 1984). Oils also contain PAHs, some species of which are known carcinogens. Some road dusts have been found to contain 8.5 μg/g of PAHs (Colwill et al, 1984).

Abrasion and corrosion

Pollutants generated by the everyday passage of traffic consist mainly of the products of abrasion from tyres and paved surfaces, corrosion of vehicles and their components, and fuel and lubricant losses. Tyre wear releases lead, zinc and hydrocarbons either in particulate form (WRc, 1993) or in larger pieces as a result of tyre failure (see Table 2.2).

Corrosion of vehicles releases pollutants such as iron, chromium, lead and zinc from rust particles and flakes of paint; their impact, as part of runoff, being largely dependent on particle size and chemical form.

The results in Tables 2.2, 2.3 and 2.4, are annual average figures obtained from a study of discharges from urban highways in Germany. Results from roads in Britain with comparable traffic densities should be similar. Note that these average figures may mask significant seasonal variation.

Table 2.2 Tyre abrasion on urban streets (after Muschack, 1990)

Type of street	Total abrasion	Lead	Chromium	Copper	Nickel	Zinc
Units	kg/(ha.a)			g/(ha.a)		
Residential way	137	60	10	12	10	35
Residential street	62	76	13	17	12	43
Distributor road	72	112	19	26	18	63
Main distributor road	109	172	29	39	27	96
Main road	127	266	44	60	42	149
Dual carriageway	120	382	64	77	61	214
Motorway	328	572	120	164	115	405

Wear of the paved surface will release various substances: bitumen and aromatic hydrocarbons; tar and emulsifiers; carbonate, metals and fine sediment, depending on the road construction techniques and materials used (see Table 2.3).

Table 2.3 Emissions from abrasion of urban streets (after Muschack, 1990)

Type of street	Total abrasion	Lead	Chromium	Copper	Nickel	Zinc
Units	kg/(ha.a)			g/(ha.a)		
Residential way	1734	177	619	88	2030	285
Residential street	2148	219	767	110	2513	352
Distributor road	3152	322	1125	161	3688	517
Main distributor road	4850	495	1731	247	5674	795
Main road	7665	782	2736	391	8968	1257
Dual carriageway	11000	1124	3927	561	12870	1804
Motorway	10000	1020	3570	510	11700	1640

Other pollutants include metal particles, especially copper and nickel released by wear of clutch and brake linings (see Table 2.4).

Table 2.4 Emissions due to abrasion of brake linings on urban streets (after Muschack, 1990)

Type of street	Total abrasion	Lead	Chromium	Copper	Nickel	Zinc
Units	kg/(ha.a)			g/(ha.a)		
Residential way	6.84	7.3	13.7	210	51.1	0.9
Residential street	8.48	9.0	17.0	260	63.3	1.7
Distributor road	12.44	16.6	24.9	381	92.9	2.4
Main distributor road	19.14	20.4	38.3	586	142.9	2.6
Main road	30.26	32.2	60.5	926	225.9	2.1
Dual carriageway	43.42	46.3	86.9	1329	324.2	5.8
Motorway	82.13	87.6	164.3	2513	613.2	11.0

Turbulence

Turbulence generated by vehicle movement can mobilise and transport particles suspended in the air; conversely deposition of particles can be induced dependent upon their size grading. Up to 40% of solids may be dispersed within a zone extending up to 5 metres either side of a motorway carriageway (Colwill et al, 1984). Particles up to 0.25 mm size can be activated by wind speeds of less than 8 km/hr (Bellinger et al, 1982).

2.3.2 Maintenance

Pollutants from maintenance operations include de-icing agents, herbicides and pesticides applied to paved surfaces, verges and the side slopes of roads. Pollutants can also arise from road re-surfacing operations.

De-icing

Sections 41 and 150 of the Highways Act of 1980 make highway authorities responsible for keeping roads safe and clearing them of snow. Thus preventative salting operations are undertaken to avoid road icing in winter. Applications of de-icing agents are the major source of dissolved solids and a significant source of suspended solids in runoff from treated roads. However, it should be noted that an earlier CIRIA project found that only major roads and 10% of others receive regular applications.

Salt (sodium chloride) is the most commonly used de-icing agent, although urea and ethylene glycol (antifreeze) are occasional alternatives which have been applied for special purposes. Rock salt for de-icing is subject to a British Standard (BS 3247) although detailed compositions can vary depending on the source. Hedley and Lockley (1975) quote an average composition of road salt supplied by ICI as follows:

Sodium chloride	90.6%	
Insoluble residue	9.2%	(mostly clay marl)
Iron	$1550 \mu g/g$	
Nickel	$12 \mu g/g$	
Lead	$8.7 \mu g/g$	
Zinc	$6 \mu g/g$	
Chromium	$4.7 \mu g/g$	
Cyanide	$5.7 \mu g/g$	

Colwill et al (1984) quote an insoluble fraction of 6%, of which 80% is grit, although it is not clear whether the grit is naturally contained or an admixture. In any event, they state that salting contributes 25% of winter suspended solids loadings (based on a rural motorway study). They also estimate that road salt contributes over 90% of total winter soluble loadings and quote bromide as another significant impurity in the salt at 230 $\mu g/g$. In addition to the contamination caused by the salt itself, its use can greatly increase corrosion rates in vehicles and metal structures, leading to increased metal deposition. Salt in solution can also create conditions that allow the release of toxic metals such as mercury from silts and sludge (WRc, 1993). It is likely that de-icing salt is a major source (at least in winter) of bromide, nickel and chromium (Hedley and Lockley, 1975) and it is suggested that it may affect the solubility and mobility of other metals, notably lead, which may precipitate more readily in the presence of sodium (Laxen and Harrison, 1977). The cyanide (as sodium ferricyanide) in road salt is added to prevent caking, and compounds containing phosphorus may also be added as rust inhibitors. In addition to its direct contribution to soluble and insoluble pollutant loadings and the increase in vehicle corrosion rates, salting may also damage the road surface (Hedley and Lockley, 1975; Bellinger et al, 1982) by either mobilising additional solids directly or rendering the surface more vulnerable to attack by freeze/thaw and road wear processes.

Colwill et al (1984) quoted 24 $g/m^2/day$ as typical for motorway application (in the south of England) with a range from 14 to 95 $g/m^2/day$. However these figures are probably out of date. Current recommendations (DOT, 1992) are for salting at 10 g/m^2 as a precaution against frost or ice (15 g/m² if the salt is wet due to open storage) and at 25-40 g/m² if freezing conditions are expected after rain or snow. Similar rates are recommended in the comparable Scottish Office code of practice.

In recent years the introduction of advanced meteorological forecasting, weather radar, thermal mapping and ice prediction techniques has resulted in a reduction in the number of occasions on which salt is used. In addition, the measurement of residual salt on the road surface now enables salt to be spread as a top-up so reducing the quantities that are used. Better vehicle design and maintenance and better driver training allows more accurate achievement of target spreading rates.

The use of salt for snow removal is also being reduced by improved highway design, including more efficient snow fences.

Salt storage dumps may themselves be a source of acute loadings if they are unprotected and located adjacent to or within the drainage area of the highway. The Trunk Road Maintenance Manual also advises on salt storage. The majority of DOT compounds have salt barns and runoff from all salt storage areas is controlled.

Because of the corrosion caused to expensive structures such as bridges, there have been trials with other de-icing agents (principally urea) on particularly vulnerable sections of highway. The use of urea can itself lead to problems in water quality (Noble and Cook, 1987) as it hydrolyses to ammonia. In un-ionised form ammonia is extremely toxic to fish and river water quality standards set a limit of 0.021 mg/l NH_3 as N for designated fisheries. The ionised form of ammonia is also toxic to fish, and both forms can subsequently exert an oxygen demand on the receiving watercourse and oxidise to nitrate, thereby adding to the problems of high nitrate levels caused by intensive farming practices. High nitrate levels encourage the formation of algal blooms in surface waters and can affect the abstraction of water for human consumption.

Roadside weed and pest control

Herbicides and pesticides are used in maintenance operations to control weeds and pests on roadsides and verges. The triazine group of herbicides, including atrazine and simazine, have been used extensively for roadside weed clearance and are more soluble and more mobile than their organo-chlorine predecessors. Combined levels of atrazine and simazine above the advocated maximum concentration of 1 $\mu g/l$ are not uncommon in watercourses near highways (Ellis, 1991b).

Herbicides are now commonly used in conjunction with mechanical cutting for roadside weed control. Their use on verges and other soft landscaping does not usually cause significant pollution of runoff. However, care must be taken to ensure that the herbicide does reach its target rather than landing on the road surface. Foster et al. (1991) found that relatively little of the applied herbicide apparently reaches the target. A more important problem is the use of herbicides on hard surfaces in urban areas. Any chemicals landing on the impermeable surfaces are likely to be carried to the drainage system where, if there are soakaways into an aquifer, there may be little scope for absorption or retention prior to infiltration. Concentrations of several hundred $\mu g/l$ have been found in first flush runoff going to highway soakaways in chalk and to surface watercourses. The situation should improve since the ban on the use of atrazine and simazine for non-agricultural purposes from August 1993. The compounds proposed to replace them have not, as yet, received the same degree of investigation. Consequently, their longer-term impact has yet to be established. Some alternatives to atrazine and simazine, e.g. diuron, are already being recorded at levels above the EC guidelines.

Highway re-surfacing and refurbishment of street furniture

Re-surfacing of highways may introduce high concentrations of both dissolved and particulate solids; however these are mainly inert. Problems with sediment blanketing of the beds of adjacent watercourses are not uncommon and can affect the local ecology. Additional loadings of fuel, oil and other hydrocarbons may result for short periods. The impact of such pollution, from fresh applications of binder with its attendant PAHs, has not been researched.

The application of road markings and the cleaning, maintenance and painting of street furniture, such as fences and lamp standards, can also be expected to generate minor amounts of contaminants from time to time.

2.3.3 Other sources of pollution

Other pollution sources include atmospheric deposition, precipitation, neighbouring agricultural activities, animal wastes and general litter.

Atmospheric deposition and precipitation

Rainwater can add its own absorbed and dissolved pollutants to the loads generated from other sources, particularly scavenged particulate matter. The presence of impermeable surfaces and highway drainage prevents natural filtration in the soil zone from removing this pollution. The quality of the rain can also exacerbate the effect of other pollution sources e.g. in coastal areas higher sodium and chloride concentrations in rainfall may increase susceptibility of vehicles to corrosion. Air pollution in industrial areas may result in enhanced solution potency, as for example with acid rain, which is essentially a dilute mixture of sulphuric and nitric acid. A road drainage study in Sweden noted rainfall pH of between 3.8 and 4.9 (Morrison et al, 1988). Similar acid levels have been recorded in east and north-east England and Scotland although the rest of Britain generally experiences more benign rainfall pH values of 5.5 - 7.5.

Snow scavenges and absorbs more atmospheric pollutants than rain and is therefore dirtier. Clearance of snow may also involve greater than normal application of de-icing materials and potentially greater damage to road surfaces. Snow dumping, particularly in Scotland, can cause significant peaks in pollution loads entering nearby watercourses during a thaw.

Dry atmospheric fallout can be responsible for large proportions of road dust. Bellinger et al (1982) quote a study in Chicago where an exceptional 70% of road dust was attributed to atmospheric fallout. This could be a significant pollution load, particularly if the particles contain soluble material, heavy metals and toxic substances. In Bellinger's case, however, the analysis revealed mainly naturally occurring material that was predominately inert. Hedley and Lockley (1975) quote figures of 116 tonnes/km^2/annum in Birmingham, but this was only 1.2% of their total for highways. The extreme difference between these two results may be partly due to atmospheric and weather patterns. However, issues such as estimation method, allocation of provenance and possible problems with sampling could have exacerbated the real differences. Nevertheless, atmospheric deposition can be a very significant proportion of the total particulate/sediment input to a highway and, since much of it is inert, inorganic and often calcareous in nature (depending upon local geology/building stone), it can prove beneficial in neutralising acid emissions, reducing heavy metal solubility and assisting in the binding of organic and other pollutants.

Agricultural activities

In temperate regions such as Britain, overland flow is only an occasional process. It generally occurs only after exceptionally heavy rainfall such as that associated with thunder storms. When it does occur it can wash soil and soluble materials on to the road surface. These solutes may include fertilisers from adjacent farmland and, in particular, herbicides from verges and cutting slopes together with natural soil particulate and humic materials.

Wind drift of chemicals sprayed on agricultural land adjacent to highways can also contribute to the pollutant load.

Roadside vehicle maintenance

Pollutants such as waste oil, hydraulic fluids, fuels (from minor spillages), antifreeze, litter, grit and organic solids arise from roadside vehicle maintenance and washing, particularly in residential areas but also in laybys, service areas and other roadside halts. These pollutants regularly contribute to the total load carried by highway runoff.

Animal wastes and litter

Urine and excrement deposited on roads and adjacent pavements by animals is a source of various bacteria and viruses and soluble and particulate contaminants with high oxygen demands. Daily excrement loadings from pets in certain urban areas can be quite significant (Muschack, 1990). Dead animals will decompose to release particulate and soluble contaminants and various bacteria, including faecal streptococci and coliforms.

Studies undertaken in continental Europe seem to indicate that pollution by such materials is made worse by mechanical sweeping or street washing. Road sweeping is intended to remove litter, faeces, leaves and so on. In reality the mechanical sweeping usually employed may do little more than break up the target into smaller particles that remain on the pavement and are more easily transported by water into drainage systems.

Deliberate disposals mostly consist of litter thrown from vehicles or deposited at the roadside, and sump oil from car maintenance. Contamination from sump oil is more likely to occur in urban areas than on country roads. Such incidents may be on the decrease due to improved recycling and disposal facilities for this type of waste. Occasionally more serious deliberate disposals occur, such as illicit toxic waste disposals into highway drainage systems.

Litter is associated principally with urban areas, though it is also deposited on country roads. Fallen leaves and grass cuttings may lie on the road surface - particularly in gutters - and decompose, or may be washed into drainage gullies. Litter will generally result in elevated levels of solids and greater consequential oxygen demand.

2.3.4 Total loadings

Information on source loadings, as opposed to runoff loadings, is relatively scarce. It is likely that many of those statistics that do exist depend, to some extent, on back-estimation from runoff as the latter is far easier to measure. However, it should be noted that most highway authorities conduct a regular programme of cleaning, sweeping and gully emptying (see Section 4.5.1) that can remove more than half of the deposited pollutants from the highway system (Hedley & Lockley, 1975). This needs to be borne in mind when attempting any back-calculations of source loadings from runoff measurements and care is needed when using different sets of figures from different studies.

There are few original British and European studies that have produced comprehensive data over at least one full year. Most of them appear to have been carried out in the 1970s and early 1980s. Table 2.5 presents results drawn from a number of different studies in the USA, Europe and Britain for a range of highway types, although most are studies of motorways with AADTs within the range 40-80 000 vehicles/day.

The results appear consistent for total annual pollutant loadings, although Hedley and Lockley's (1975) study of the Aston Expressway (A38(M)) in Birmingham gives some very high results. This may be related to the industrial nature of the region, although atmospheric depositions only accounted for 1.2% of the total loadings. A possible explanation might be that these high results were caused by extensive winter de-icing operations that produced chloride ion concentrations often 1000 times those normally recorded, whilst other pollutants were typically 5-10 times their concentrations for the rest of the year. Such unusually high loadings during the winter period may have unduly influenced the overall annual pollutant loadings quoted in Table 2.5.

Vehicles /Day (AADT, thousands)	Highway type	Suspended solids	COD	Oil	NH₄-N	PO₄-P	Pb	Zn	Cu	Study area	Study dates	Source of data
< 5	Residential	2218					0.30	0.40	0.39	Germany	1986-89	Muschack (1990)
	Residential	3640								UK	1991-92	Butler (1993)
5 - 15	Urban roads	4978					0.68	0.89	0.87	Germany	1986-89	Muschack (1990)
	Various urban	7289								UK	1991-92	Butler (1993)
< 30	Various urban	550	245	-	3.6	1.0	2.2	1.4	0.5	USA	1976-77	US-FHA (1981)
41	Rural motorway	873	672	43	4.6	1.6	1.3	2.3	0.6	Germany	1978-81	Stotz (1987)
47	Rural motorway	848	557	27	3.2	1.5	1.1	2.9	0.5	Germany	1978-81	Stotz (1987)
> 50	Various motorways	1930	139[1]	168[2]	3.3	-	3.1	4.6	1.2	UK	1980s	Ellis (1991b)
	Urban motorways	10410	-	-	-	-	1.68	2.06	3.19	Germany	1986-89	Muschack (1990)
> 60	Urban motorway	6289	-	-	-	-	12.9	19.0	3.8	UK	1973-74	Hedley and Lockley (1975)
65	Rural motorway	1000	-	85	-	-	3.0	5.8	-	UK	1980-81	Colwill et al (1984)
	Motorway	770	-	-	-	-	2.4	1.9	-	Switzerland	1970s	Dauber et al (1978)
	Motorway	650	-	-	-	-	1.2	1.9	-	France	1970s	Cathelain et al (1981)

[1] BOD₅
[2] Total hydrocarbons

Table 2.5 Annual pollutant loadings from highway traffic (units = kg/ha/a)

2.3.5 Quality of runoff

Runoff quality depends upon a number of factors. These include:

- road and traffic profiles and characteristics
- weather and any local micro-climate factors that influence incident precipitation
- evaporation rates and the antecedent dry period
- the road drainage system, its characteristics and serviceability.

A worst case is a short duration intense summer storm, which generates a small total quantity of rainfall after a long antecedent dry period. In this case pollutants that have accumulated during the dry period are mobilised by the high intensity rainfall and discharged into a stream that is probably at a low flow. More extensive rainfall would have less impact due to the greater dilution of pollutants in the runoff and the higher flows in the receiving stream.

Many researchers have reported first flush effects, in which the first runoff generated by a storm carries an unusually high contaminant loading as pollutants that have accumulated during the antecedent dry period are removed. The pollutants may have accumulated on the road surface or - in the case of sediment particles - may have accumulated in the drainage system after previous runoff events; this may cause a double peak of suspended sediment as the detritus lodged in the system is removed first, followed by material from the road surface itself.

Bellinger et al (1982) reported this effect and a variety of other first flush phenomena. However, they did not find consistent behaviour and remarked that the effect was clearly a function of variables, such as the pattern of rainfall on the catchment, the period of travel and the distribution of solid material over the catchment. Harrison and Wilson (1985a) found that although the first flush can be significant, the importance of its effect varied from storm to storm. They observed that removal of soluble material is not a function of rainfall intensity as even light rains removed soluble deposits that had accumulated on the road since the last runoff event. Sediment particles, on the other hand, may remain on the road surface until the intensity of rainfall is sufficient both to mobilise them and transport them from the road surface. The impact of raindrops may be significant in loosening some particles. Storm events are more likely to result in particles and oils evading interception in the road drainage system because of the higher energies and loadings involved. Other contaminants associated with the particles will be removed with them. This means that soluble and insoluble forms of the same contaminant may peak at different times in the discharge from a single storm.

The length of the antecedent dry period is significant in various ways for different contaminants. Soluble species deposited by traffic or atmospheric fall out, for example, will collect more or less in proportion to the length of time since the last runoff event. Those deposited by irregular events, e.g. de-icing salt, will not accumulate so predictably and may or may not form a major component of first flush runoff. Insoluble materials and those associated with them, including both heavy metals and hydrocarbons, will collect until there is a storm of sufficient intensity to remove them; the intensity required being dependent on the particle size. Thus, for insoluble and adsorbed materials, antecedent dry period refers to the period since the last rain of sufficient intensity to move them. As absorbed contaminants may also attach preferentially onto smaller particles, their movement follows an especially complex pattern.

Pollutant	Rainfall	Motorway I	Motorway II (avg.)	Urban I	Urban II	Urban III	Rural
Electrical conductivity (μS/cm)	8-80	25-22 000	25-18 000 (550)	6-20 000	-	-	-
Total solids	18-24	<15-12 860	110-5700 (261)	145-21 640	11-40	-	-
Total dissolved solids	-	12-12 560	-	66-3050	-	-	-
Total volatile solids	-	<20-940	-	12-1600	-	-	-
Volatile suspended solids	-	-	-	12-1500	-	20-78	6-25
Total suspended solids	2-13	18-3430	-	2-11 300	-	68-295	12-135
Oil/hydrocarbons	-	6-40	8-400 (28)	0-400	3-31	-	-
COD	2.5-32	36-575	-	5-3100	-	57-227	28-85
Chloride	1-11	1-6714	159-2174 (386)	4-17 000	4-27	-	-
Bromide	-	0.05-6.0	-	0.02-6.0	-	-	-
Total lead	0.000024-0.01	0.1-8.0	0.34-2.4 (0.96)	0.01-14.5	0.01-0.15	0.1-1.5	0.024-0.27
Total zinc	0.00002-0.05	0.12-4.0	0.17-3.6 (0.41)	1.0-15	0.02-1.9	0.19-0.56	0.035-0.18
Total cadmium	0.000013-0.000056	<0.003-0.1	-	0.002-0.4	-	-	-
Total copper	0.00006-0.0005	0.007-0.03	0.05-0.69 (0.15)	0.007-2.5	0.010-0.12	0.025-0.12	0.01-0.05
Total chromium	0.000023-0.00008	0.018-0.085	-	0.018-0.27	-	-	-
Total nickel	-	0.036-1.55	-	0.02-1.5	-	-	-
Total organic carbon	1-18	-	-	5-120	-	8-74	3-17
Nitrate & nitrite	0.01-5.0	-	-	0.3-6.9	-	0.4-1.5	0.2-0.9
Total nitrogen	0.5-9.9	-	-	0.2-14	0.2-1.0	1.0-3.2	0.3-2.2
Total phosphorous	0.001-0.35	-	-	0.3-4.4	-	0.2-1.0	0.1-0.5
BOD	1-15	-	12-32 (24)	25-700	8-25	-	-

N.B. The average concentration is typically three times the lowest concentration.

Table 2.6 Reported ranges of pollutant concentrations found in various locations (mg/l unless stated)

(Colwill et al, 1985; Strecker et al, 1990)

The quality of runoff is also influenced by the form of potential pollutants. Fine particles may be more easily mobilised into suspension, although Bellinger et al (1982) state that mobilisation is independent of size within the range 0.01-1 mm. Surface roughness may also restrict suspension by providing refuges for particles. These influences are crucial because the majority of pollutants (the main exception is road salt) are carried in or associated with particles. Collins and Ridgeway (1980) found that, from an analysis of street dust, fines <63μm constituted only 6% of the total. However, this small component accounted for the following pollutant loads:

- 25% of the oxygen demand
- up to 50% of all nutrients
- over 50% of heavy metals
- nearly 75% of pesticides.

Hewitt and Rashed (1992) give figures for a rural highway showing that the particulate fraction contained over 90% of the total inorganic lead, 70% of the copper and 56% of the cadmium. Most oils and hydrocarbons are also associated with particulate matter. It is clear, therefore, that the majority of perennial pollutants are associated with the particulate matter in the runoff.

A large number of studies have been conducted on the quality of runoff from different road types. Table 2.6 provides examples of the ranges of figures recorded (from Colwill et al, 1984; Strecker et al, 1990). The maximum COD values recorded for highway and urban runoff are greater than values obtained for combined sewer overflows (CSOs) (Strecker et al, 1990). However, typical values are one order of magnitude lower than this.

Colwill et al (1984) reported on a study of the quality of runoff from a short length of the M1 motorway in Bedfordshire. Quality variations between November 1980 and October 1981 were reported and substantial differences were found between winter and summer, mainly due to the use of de-icing salt in winter. Similar studies, Bellinger et al. (1982) on a much longer length of the M55 north of Preston and Hedley and Lockley (1975) on the Aston Expressway, reported similar results.

Few references to microbial contamination have been found in the European literature surveyed. American Federal Highway Administration Reports (FHA, 1981) provide some interesting results from the monitoring of the usual pathogenic indicator bacteria (i.e. total coliform, faecal coliform, and faecal streptococcus) in highway runoff. High concentrations of all indicator bacteria were found in all runoff samples from a wide variety of US sampling sites. Typical values were 10^4 - 10^6 cols/100ml, rarely dropping below 10^3 cols/100ml. First flush effects were not always apparent, with high results often maintained throughout a storm runoff event. Salmonella organisms were often found when included in the analysis. The FHA generally found that animal and bird droppings were the major source, followed by soil and dust materials and animal deaths.

2.4 ACCIDENTAL SPILLAGES

2.4.1 Types of spillage

Although probably minor in terms of overall loadings, spillages resulting from individual accidents are potentially the most serious source of contaminants associated with highways. Accidental spillages can range from minor losses of fuel from vehicles to major losses from fractured tanker vehicles, but their effects can be serious because of the unpredictable nature of the materials involved. The range of potentially polluting materials transported by road in the United Kingdom is too large to be listed in this report. It is generally acknowledged that spillages of solid materials, whether in loose or bagged form, from inadequately secured loads present a significant potential threat of pollution in Britain where many minor accidents typically involve partial load shedding or spillage. Bagged chemicals, paints, food products and waste materials of all types are some of the more serious pollutants concerned.

Liquids, which are carried in large quantities, also present a high potential for serious pollution following accidental spillage. Such materials include:

- petrol, diesel fuel, oils, other liquid hydrocarbons and chemicals
- acids and caustic solutions
- toxic wastes
- inert slurries
- sewage sludge
- products that can cause high biological loadings e.g. sugar and dairy products.

The pollution potential of these materials varies considerably. Some (petrol and oil) have already been discussed in the context of leakage and emission. Slurries and sludge, if inert, will add to the suspended solids loadings. Waste sludge (including sewage sludge) can have a very high oxygen demand and can also include toxic metals. Foods, sugars and dairy products such as milk have very high oxygen demands that could potentially render any receiving watercourse anoxic. Acids and alkaline solutions are generally treated by neutralising solutions and powders at the spillage site, but residual solutions of low or high pH may result if neutralising quantities are not carefully controlled. Acids will mobilise into solution many heavy metals both from the road surface and from accumulated dust and grit.

Other transported liquids pose less of a problem because they solidify or evaporate under ambient conditions. These include some chemicals that are transported hot as well as cryogenic gases such as liquified oxygen and nitrogen.

Legislation controls HGVs and the loads that they carry. The Construction and Use Regulations are intended to prevent overloading, whilst the Carriage of Dangerous Substances Regulations specify safe loading and filling ratios and requirements for hazard warning panels. The Department of Transport also issues guidance in the form of a Code of Practice on Safety of Loads on Vehicles. Nevertheless, it has been suggested that stricter enforcement of existing legislation to control HGVs and the loads that they carry may be required; this is recommended for future study (Section 7.1). In particular, stricter enforcement of the Construction and Use Regulations to prevent overloading, more thorough checks on methods of securing loads and scrutiny of bills of lading combined with more frequent and compulsory weighbridge checks as is practised in North America, may reduce the number of incidents. Legislation, requiring the improvement of containers for all road transported materials, may result in a long-term reduction in the occurrence of accidental spillages.

The potential pollution problem will always be exacerbated if the spillage occurs during a period of rainfall. Solid or liquid materials, which might otherwise remain relatively immobile, allowing their physical removal, can be dispersed by highway runoff.

It is still a common practice for fire services attending an accident to carry out water flooding of the affected area to avoid the risk of fire or explosion. This inevitably leads to increased pollutant mobility through the drainage system. However, the emergency services are aware of the risk of pollution and endeavour to seal off entries to drains when such a risk is apparent. More equipment is being made available to them to enable spilt pollutants to be contained and the HAZCHEM Scheme is currently undergoing revision to take more account of environmental impact and to recommend containment of spillage for many more substances.

Less dramatic but more frequent accidental spillages will occur when vehicle fuel tanks, crankcases, radiators and gearboxes are damaged and leak following accidents. Accidents may also involve the contamination of the highway with debris, such as broken glass, rubber, metal and paint flakes from damaged vehicles and structures, and with the products of combustion.

2.4.2 Number of incidents

The general public believes that there is a great hazard from accidental spillage. About 120 incidents per year have been reported on roads in the Greater London area, the majority being minor petrol spillages. The Chemical Industries Association (CIA) estimates 200-300 chemical spillages per year throughout Britain, although presumably this excludes some polluting products such as dairy, food and waste products. The majority of the 200-300 estimated spillages involved only a few litres. The Health and Safety Executive has laid down reporting requirements for accidents under its RIDDOR Regulations and the Home Office has statistics compiled by its Fire Services Inspectorate. Statute also requires that pollution incident registers are compiled by water regulatory authorities.

However, there are no published national statistics for accidental traffic spillage (Bickmore and Dutton, 1984). Few registers distinguish between pollution from traffic accident spillages and that arising from unauthorised releases from installations. In some instances the large volume of data recorded prevents significant analysis. Analysis of currently available statistics shows that such incidents are relatively rare. The national average for notifiable accidents on British motorways is 1 : 8.33 million vehicle km.

The questionnaire survey requested details of the availability and standard of information recorded in respect of pollution incidents. Although every water regulatory authority has to maintain a pollution incident register, only one Authority's register was formatted so that highway/traffic accident incidents could be identified and collated. This is the database operated by Thames Region of the NRA (inherited from the former Thames Water Authority), in which it is possible to distinguish between pollution incidents arising from road transport and those arising from other sources. A study of the database for the last five years reveals the statistics summarised in Table 2.7 below:

Table 2.7 Extracted data from NRA (Thames Region) pollution incident register

Year	1988	1989	1990	1991	1992	1993 (to July)
Number of incidents (total reported)	2811	3613	3444	3417	3598	1979
Road incidents reported	125	177	175	136	212	104
Road incidents where pollution substantiated	88	63	64	42	58	36
Road incidents category 1 & 2	26	21	5	1	3	4
Road incidents category 1 & 2 substantiated	26	14	5	1	3	3

Notes to Table 2.7:

- the NRA classifies pollution incidents into three categories 1 (Major), 2 (Significant) and 3 (Minor). The drop in incidents of categories 1 and 2 between 1989 and 1990 probably reflects the change in classification criteria with the change from Thames Water Authority to the NRA, rather than a real change in occurrences of pollution
- of the 52 substantiated category 1 and 2 pollution incidents arising from roads, 29 involved oil or petrol
- major spillages from road transport vehicles are included but the majority were from vehicle fuel tanks following some accident or mechanical failure
- other pollutants registered included: various chemicals, dyestuffs, pesticides,

- fungicides, beer, paint, adhesives, sewage sludge, milk and offal
- the incidents include several that originated from vehicles or trailers that were parked on or near roads and that lost some or all of their loads as a result of vandalism or some other unidentified cause
- two incidents were described as originating from urban runoff, in which first flush runoff after heavy summer rain following a dry period was believed to have caused a sudden drop in dissolved oxygen in the receiving streams
- two incidents involved soakaways and two occurred at places where drainage appears to go into the ground with no watercourse or infiltration structure involved
- the remaining incidents involved surface watercourses.

There are no publications dealing with the risk assessment or prediction of such incidents in relationship to identifiable road features such as roundabouts, junctions or accident blackspots. One very limited study, which was conducted in support of an evaluation of aquifer vulnerability (Price et al, 1989), suggested that the risk of an accident involving an HGV occurring at motorway interchanges is over twice that associated with the sections between interchanges. A tendency for accident rates to be significantly higher in the 12 months after opening a new interchange was also discovered. Unfortunately the figures used in this study do not include classified traffic volumes upon which to base a detailed analysis, nor do they allow for analysis of pollution potential.

It is possible, from existing data registers and reports such as the Health and Safety Commission Advisory Committee's report on Dangerous Substances of 1991, to derive meaningful risk assessments of the potential for accidental spillages and even to categorise them in terms of probability for different classes of road or road alignments. To undertake such a task is beyond the scope of this report. However, such an exercise could be of great value.

2.5 SCALE OF POLLUTION PROBLEMS

Although the discharge of highway drainage to surface and groundwaters occurs throughout the country, the evidence of significant pollution from this source is largely anecdotal.

2.5.1 Pollution of surface water

The nature of the receiving watercourse obviously has a major effect on the impact resulting from discharges of highway runoff. The dilution available for the discharge, the quality of the water in the receiving watercourse and the catchment characteristics of the watercourse (slope, geology, land use etc) will all determine its response to drainage inputs. Storm events will often generate natural increased sediment loadings within the watercourse, which may mask additional loading from highway runoff.

Potential impacts include:

- blanketing of stream beds with sediment, affecting habitats and food supply to bottom feeders
- accumulation of contaminants in sediments deposited at point of discharge with consequent problems for final disposal
- problems of disposal of dredgings from the bed of receiving waters, as such material is now defined as a controlled waste under the Environmental Protection Act 1990
- increased turbidity and suspended solids, affecting filter feeders and gill efficiency, and light penetration for photosynthesis; this may be exacerbated by turbulence at the outfall itself (Ellis, 1991b).
- depletion of dissolved oxygen due to enhanced chemical and biochemical oxygen demands
- algal growth and eutrophication due to increased levels of nutrients

- direct toxic effects on stream biota due to soluble pollutants or ingestion of particulate pollutants
- taste, tainting or toxic effects resulting in closure of downstream water supply intakes
- smell
- loss of visual amenity.

Although the cause might be short-lived, the impact itself may have a longer duration because of the time needed for recovery of the stream ecosystem.

A number of studies have conducted both field and laboratory measurements on the biological impact of highway runoff as a single polluting entity. Lord (1987) conducted "worst case" laboratory experiments by exposing a variety of typical stream flora and fauna to undiluted highway runoff from roads with a variety of AADT flows. His results confirmed the general opinion of a number of workers in this field that at AADTs <15 000 veh/day there are virtually no noticeable effects, while at AADTs >15 000 to <30 000 only minor impacts were observed (Maestri et al, 1988; Ellis, 1991b). Thus only heavily trafficked main roads, urban distributors, trunk roads and motorways may require specific interception and retention drainage measures.

Salt from winter maintenance has also been recorded as having a polluting impact. Elevated chloride concentrations are rare in Britain and are likely only to occur as a peak after a de-icing operation; such peaks would very quickly become diluted. Bellinger et al (1982) report a doubling of the dissolved ions of sodium (25 to 47.5 mg/l) and chloride (28.5 to 51 mg/l) in the Woodplumpton Brook downstream of the nearby M55 in Lancashire.

Freshwater fish vary in their tolerance to salinity. Perch and rainbow trout can withstand up to 14 000 mg/l, whereas chloride levels around the drinking water maximum limit of 400 mg/l could stress a variety of less robust freshwater fish and some invertebrates (Bickmore and Dutton, 1984). The immediate area around an outfall can show seasonal reduction in non-mobile sensitive species such as flatworms.

Drinking water abstractions might be shut down temporarily because of increases in salt levels. For example, Brownsover and Thornton reservoirs, near to the M1 and M6 in Warwickshire, have experienced occasional increases in chloride concentration in winter from 40-80 mg/l to 200 mg/l (Bickmore and Dutton, 1984). In this case shut down of the reservoir abstraction was not required. Farm animals are more tolerant of salt than man and will not therefore be affected by salinity peaks in watercourses.

Dobson (1991) presents a well referenced work on the effect of de-icing salts on roadside vegetation and discusses cost-effective alternatives to current practice.

The questionnaire survey for this project indicated a case in central Scotland where highway runoff was possibly the cause of the shut-down of a village's water abstraction. First-flush runoff repeatedly caused pollution (visible oil sheen) of a watercourse. The drainage of the trunk road passing through the village is connected into one ancient catchpit, which discharges into the watercourse. All the reported incidents occurred after summer storms following a considerable antecedent dry period. Each event was logged but no analyses for specific pollutants were carried out.

2.5.2 Pollution of groundwater

Pollution of groundwater is less immediately apparent than pollution of a surface watercourse. There are no immediate effects on fauna and flora and there is unlikely to be any public awareness or media interest. Nevertheless it may represent a serious problem for the following reasons:

1. Serious pollution of an aquifer will usually persist for a considerable time because of the slow rate of percolation through the aquifer.

2. The pollution may remain unnoticed for a considerable time so that no action is taken to prevent the polluting discharge, so increasing the pollution problem.
3. The natural processes that break down many contaminants in surface waters appear to operate much more slowly, if at all, in the subsurface environment. This is a particular concern for trace pollutants such as herbicides. Work should be carried out to determine their fate in groundwater.
4. It is almost impossible to get access to the pollutant in the aquifer to remove it or treat it.

One of the major advantages of a groundwater source for potable water is that it can be abstracted locally with little or no treatment other than disinfection. Having to treat groundwater to remove the products of pollution, which could have been prevented, would be an expensive departure from current practice. Permitting groundwater pollution is in any case contrary to national policy and to the EC Directive on Groundwater Protection. It could also have other impacts as polluted groundwater could eventually find its way into surface watercourses where, as base flow, it could cause pollution over a long time-scale.

It is clear from previous work that, under certain circumstances, a contaminant entering a highway soakaway could reach a public supply well (Price et al. 1989, 1992). However, little information relating to groundwater pollution is in the public domain. In general an organisation that thinks it may have polluted an aquifer will be reluctant to publicise the fact. Water supply undertakings, whose sources have become polluted, may also be reluctant to release information, for fear of alarming customers. Information given to the water regulatory authorities is in the public domain but this fact can discourage people from volunteering it. Runoff from highways is a potential threat to underground water resources although in Britain there seems to be little evidence of pollution of groundwater from discharges from highway drainage. The available evidence is presented below.

Routine discharges

Given the high chloride levels associated with winter runoff from motorways, chloride might be expected to be a good indicator of pollution. Chloride is a very conservative substance, and analysis for it is performed regularly and with reasonable accuracy. Therefore elevated chloride levels might reasonably be expected to have been detected in groundwater in areas where major roads drain to aquifers. The M1 Motorway in Hertfordshire is more than 30 years old and drains to solution features in the chalk near to several public supply abstraction wells and yet Medhurst (1992) reports no detected increase in chloride levels at any of the sources.

Similarly, the section of the M55 described by Bellinger et al (1982) drains to a stream flowing across a sandstone outcrop. Samples taken from a borehole in the sandstone since opening of the motorway have only shown slight increases in chloride levels (from 20 to 22 mg/l). This slight increase is less than the error band of the test results.

In another case a public supply was affected by salt attributed to runoff from a county highway depot's salt store. Even here the increase in chlorides was only a few mg/l (Medhurst, 1992).

In contrast, there are well-documented examples of the effects of road de-icing on the quality of groundwater in North America. For example, Todd (1980) quotes a case where roadway de-icing led to an increase in chloride concentration at Burlington, Massachusetts, from an insignificant count to 250 mg/l over a period of 25 years. This led to a ban in 1970 on the use of de-icing chemicals on the streets of the city. Saleem (1977) also reported much higher values of chloride arising from the application of de-icing salt near Chicago. More recently, Howard and Beck (1993) give examples of groundwater contamination by de-icing salts in Ontario, with chloride levels up to 20 times natural background levels.

One possible reason why such significant increases have not been reported in British groundwaters is that the porous matrix of the Chalk aquifer and the high storage of porous sandstone aquifers provides a high retention capacity and this will limit the area of impact of local point source contamination. This is in contrast to the diffuse pollution attributed to the effects of changed agricultural practices, giving rise to increased nitrate levels appearing in groundwater abstracted for public supply.

Accidental discharges

Problems of pollution from accidental discharges were investigated in the questionnaire survey carried out for this project.

Only one case was revealed where groundwater was known or thought to have been polluted by runoff from roads. In this case a borehole in Cambridgeshire has experienced short periods of hydrocarbon pollution that seem to be related to runoff from adjacent trunk roads. The pollutant is apparently weathered oil at concentrations of 500-600 μg/l. The aromatic components had apparently been removed by weathering, so the oil imparted no taste to the water.

Other cases have reported relatively high levels of industrial solvents in groundwaters, some of them leading to loss of public supply sources (e.g. Chilton et al, 1990: Revitt et al, 1990; Eastwood et al, 1991; Lawrence and Foster, 1991). They highlight a problem in that many instances of contamination of groundwater have been identified or investigated some years after they occurred. After such a lapse of time it may be impossible to discover the true source of the contamination. It may have been leakage that entered the ground directly from a tank or buried pipe, or the result of soakaway drainage taking one or two concentrated charges of pollutant, or a long-term effect from drainage carrying a persistent pollutant loading. The information publicly available suggests that leaks from static installations may be the greatest problem (Tester and Harker, 1981).

Tester and Harker (1981) suggest that significant quantities of pollutants, particularly hydrocarbons, could be present in aquifers as a result of leaks and spills on highways. Hydrocarbon fuels, for example, except for a small soluble fraction, can 'float' on the water table and remain undetected even when they are close to production sources. However, as Tester and Harker point out, many of these substances cause an objectionable taste or odour in water at very low concentrations and it therefore seems unlikely that significant damage to public supplies would have gone undetected unless the aromatic components had been removed by weathering or some other natural process of decay.

One incident that was recorded and investigated involved a spill of approximately 9 000 litres of nail varnish remover from a tanker involved in an accident on a motorway where the drainage is directed into a natural depression in the chalk. After the accident a nearby pumping station was closed as a precaution. An investigation was carried out by WRc, which revealed that the solvent apparently only penetrated to about 1-2 m below the surface. The bulk of the liquid had been adsorbed by a layer of silty/clay material and organic residue in the floor of the depression. The presence of the solvent was easily detected by its smell; however it was never detected at the abstraction bore, which was subsequently returned to service.

2.5.3 Aquifer recharge

It is clear from previous work that, under certain circumstances, a contaminant entering a highway soakaway could reach a public supply well (Price et al. 1989, 1992). However, highway runoff can be a significant and valuable source of recharge to aquifers. It may, therefore, be desirable to permit highway discharge to aquifers in order to provide recharge, although with adequate precautions against pollution. This additional recharge may even reduce the risk of pollution of an aquifer by keeping the water table high and preventing polluted surface water being drawn in from rivers (Figure 2.2).

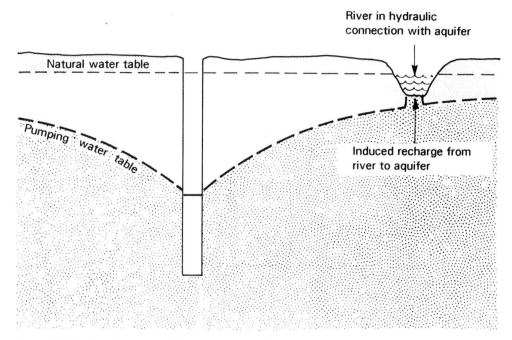

Figure 2.2 *Induced recharge from river to aquifer*

Rain falling on a natural or cultivated soil will be disposed of in three ways:

- evaporation and transpiration
- lateral flow to a drain or stream
- aquifer recharge.

Not all the precipitation falling on paved surfaces flows to drains. The first water falling on a dry surface wets the material and forms a film on the surface. This film, which can only be removed by evaporation, is termed the wetting loss. For most surface materials this is typically less than 0.5 mm (Van de Ven et al, 1992).

Once wetting is complete - usually within a few minutes - runoff will begin to be generated, but water will be retained as depression storage in hollows in the pavement. The water retained as wetting loss and depression storage can be referred to in combination as the initial loss; Van de Ven et al (1992) reported figures of less than 1.1 mm for the initial loss. However, it is not clear whether these take into account the losses that might be expected from the irregular surfaces of British roads. For example Harrison and Wilson (1985), in their study of a stretch of the M6 in Cumbria, used a model that assumed surface retention of water equivalent to 2.2 mm of rainfall in an event. They found that this model underestimated the amount of retention and commented that a dry road surface can apparently retain a lot of water because of the presence of cracks that have to be filled before runoff is generated. This appears to be an aspect that merits further study.

The dry conditions that have been experienced in Britain in the late 80s and early 90s, with high soil-moisture deficits persisting for much of the year, have meant that a single rainfall event of say 10 mm would frequently not cause recharge through grassland. The same event would theoretically, after deduction of the initial loss, result in about 9 mm of runoff from a smooth highway.

For much of eastern England the mean annual rainfall is in excess of 600 mm, but hydrologically effective rainfall is only about 150 mm/year, and in recent years has been much less. The NRA estimate that in a 1 in 50 year drought it can fall to about 40 mm in their Anglian Region. On the basis of the figures above, drainage from highways could collect more than 500 mm of the rain falling on them in a normal year. Highways therefore represent significant potential for recharge to groundwater. This can be demonstrated by the following theoretical calculation.

There are about 273 000 km of public roads in England, of which motorways and trunk roads account for only about 11 000 km. On the assumption that the proportion of public roads on the Chalk is in the same proportion as the area of land covered by the Chalk, a first approximation indicates some 52 000 km of road in chalk areas. The average width of these roads is estimated as 8 m. The calculation then gives 416 km² of impermeable highway in chalk areas.

The runoff from 500 mm of rainfall from that area would yield 2.1×10^8 m³ of water; this represents about five per cent of the average annual recharge into the Chalk, or the equivalent of 10 mm of infiltration into the whole of the Chalk outcrop. Although this may not sound a great deal for most years, it could theoretically represent 25 per cent of the recharge to the Chalk of East Anglia in a drought year (Price, in press). Other aquifers in the relatively drier parts of Britain will similarly benefit from concentrated recharge via soakaways. The Lincolnshire Limestone, which outcrops in a belt running north-south through Humberside and Lincolnshire, is an example.

After a period during which there was greatly reduced recharge into British aquifers for four consecutive years, there is a need to reconcile the issue of protecting groundwater quality with that of protecting and/or augmenting groundwater resources. Even if it were possible, in the interests of protecting water quality, to avoid disposing of highway runoff into controlled waters, the need to conserve resources suggests that this source of water should be made available for water supply or diverted into strategic storage wherever possible. There is therefore need for research into the fate and transport of pollutants in soakaway systems and in groundwater to determine the true scale of the pollution risks.

3 Standards, legislation and practice

3.1 INTRODUCTION AND OVERVIEW

3.1.1 Introduction

This Section examines the legal issues which arise in relation to the pollution of surface waters and groundwater due to runoff from roads. Specifically, the objectives of the Section are to provide a concise practical description of the legal and administrative practices in relation to roads and water pollution control in England and Wales and Scotland current at 1 January 1994. In so doing it encompasses a discussion of the implications of relevant European Community environmental law within these jurisdictions.

Whilst the precise technical issues surrounding water pollution from roads are considered in detail elsewhere in this report, a working concept of the kinds of activity which give rise to pollution from them is necessary from the outset. At risk of over-simplification, there are essentially three kinds of occurrence that may give rise to pollution originating from a road: first, rainwater runoff of traffic generated residues of oil and other substances which accumulate on roads in normal use; second, the runoff of matter used during the maintenance of roads for various purposes, such as chemicals applied for de-icing or herbicidal purposes; and third, substances transported by road which may enter drainage systems as a consequence of spillage arising from traffic accidents or the ensuing clean-up operations (see Royal Commission on Environmental Pollution, 16th Report, *Freshwater Quality* (1992) Cm.1966 paras. 7.78 to 7.80). It must be stressed, however, that this list is very generally formulated and may usefully be supplemented by additional categories relating to wear and tear and pollution from miscellaneous sources (see Section 2.1.1).

As has been stated, this Section describes the law relating to water pollution from road drainage operative in both England and Wales and Scotland. The approach taken is to progress from topic to topic, first examining the law of England and Wales and following this by a discussion of the corresponding law in Scotland. To a large extent the general state of the law in the two jurisdictions is similar, and for this reason the Scottish law tends to be reviewed more concisely with cross referencing to previous sections where further discussion of the legal principles is to be found. However, where significant differences exist, a more detailed account of the legal and administrative principles, with emphasis upon points of contrast between the two jurisdictions, has been incorporated.

3.1.2 Overview

The potential problem of water pollution as a result of road runoff in England and Wales and Scotland is an issue involving a range of public authorities with responsibilities for the protection of the aquatic environment and for roads. In England and Wales the main body responsible for water protection is the National Rivers Authority (NRA), which is possessed of a wide range of functions relating to environmental protection including water pollution control, water resources protection and various ancillary matters such as planning. In Scotland more limited powers are entrusted to river purification authorities, though these include key duties in relation to water pollution control.

With respect to highway responsibilities, in England and Wales these are divided between the appropriate Secretary of State and the local highway authorities, whilst in Scotland corresponding duties relating to roads are allocated to the Secretary of State for Scotland and the local roads authorities. Although various differences exist in matters of detail, broad similarities are to be found between the powers and duties given to bodies in England and Wales and their Scottish counterparts. Significantly, duties in relation to maintenance give rise to a range of potentially polluting activities and, in each instance, legal mechanisms are provided for in order to prevent water pollution.

The maintenance of water quality is increasingly a matter determined by European Community law, and a number of key directives are relevant to the study. Of particular significance are Community directives concerned with dangerous substances and groundwater quality. In each instance Community directives must be comprehensively implemented within member states and consequent duties arise for the United Kingdom Government to ensure that national legislation has this effect. Particular consequences of this are provisions for the specification of statutory water quality objectives for waters of different kinds. Parallel legislation is provided for in England and Wales and Scotland to impose water quality objectives and to require regulatory authorities to use their legal powers to ensure that objectives are met and maintained.

In particular instances it is possible that the entry of polluting matter originating from a highway and passing into a watercourse or groundwater might constitute a criminal offence, and the details of this offence require close consideration in relation to the context at issue. The meaning of the statutory formula of "causing or knowingly permitting" water pollution is considered alongside the various defences and statutory authorisations which are provided for in relation to this offence.

Of special significance to the study is the legal position of highway drains in water pollution law. Whilst a general defence to the water pollution offence is provided for in respect of matter passing through a highway drain, this defence can be lost where the regulatory authority imposes a prohibition upon discharges through the drain or, under certain circumstances, imposes a discharge consent in relation to the drain. Although the legal details differ somewhat, the same basic principles governing water pollution from highway runoff are to be discerned in the laws of England and Wales and Scotland.

Legal issues relating to the protection of water resources are also relevant in respect of pollution originating from highways, and issues of both criminal and civil law may be pertinent. Significant differences exist between the statutory provisions which are made for the protection of water in England and Wales and in Scotland. However, in both jurisdictions important aspects of civil law need to be considered in determining the extent of private rights and duties in relation to the pollution of watercourses and water contained in underground strata. Particularly where water is to be used as a source of supply, requiring relatively high standards of purity, the potential for civil liability is a significant consideration.

3.2 LEGAL POWERS AND DUTIES

3.2.1 The National Rivers Authority in England and Wales

The Legal Status of the National Rivers Authority

The NRA is the principal body with responsibility for the protection of the aquatic environment in England and Wales. The NRA was established under the Water Act 1989, at the same time as the water utility functions of water supply and sewage treatment were transferred to the privatised water services companies. Since the codification of water legislation in 1991, the legal status of the NRA has been provided for under the Water Resources Act 1991. This Act commences with the statement that there is to be a body corporate known as the NRA for the purpose of carrying out specified functions attributed to it. The "principal functions" of the NRA concern water resources, water pollution, flood defence and land drainage, fisheries, functions as a navigation authority, harbour authority or conservancy authority and other functions which may be statutorily assigned to it {s.2(1) WRA 1991}. In respect of the control of pollution, the NRA has extensive powers and duties in relation to controlled waters, including the responsibility to monitor water quality, to determine applications for discharge consents and to pursue legal proceedings in relation to pollution incidents.

Environmental duties

Each of the particular functions exercised by the NRA, including that of water pollution control, is subject to certain overriding environmental obligations. Hence, it is stated to be the duty of the NRA, to such an extent as it considers desirable, generally to promote the conservation and enhancement of the natural beauty and amenity of inland and coastal waters and of land associated with such waters, and the conservation of flora and fauna which are dependent on an aquatic environment {s.2(2) WRA 1991}.

Environmental obligations also arise under the "General Environmental Duty" of the NRA which requires it, when formulating or considering any proposals, to exercise any power so as to further the conservation and enhancement of natural beauty and the conservation of flora, fauna and geological or physiographical features of special interest {s.16(1) WRA 1991}. Similar environmental duties are also imposed upon the Secretary of State, the Minister of Agriculture, Fisheries and Food and water and sewerage undertakers {s.3 WRA 1991}. Also, as will be discussed later, the NRA is subject to a general duty to exercise its powers with particular regard to the duties imposed upon the water industry ({s.15 WRA 1991} and see 3.5.1 below).

Some guidance as to the manner in which the General Environmental Duty will be exercised in particular contexts is provided for by a ministerial power to approve codes of practice relating to the Duty {s.18 WRA 1991}. This power has been exercised through the enactment of the Water and Sewerage (Conservation and Recreation) (Code of Practice) Order 1989 {SI 1989 No.1152} approving the *Code of Practice on Conservation, Access and Recreation* (published in 1990, revised 1991).

Pollution control and planning

Whilst the criminal law relating to water pollution is considered below (see Section 3.4), it is convenient here to describe the general role of the NRA in seeking to prevent developments which have a damaging effect upon the aquatic environment through the operation of planning law under the Town and Country Planning Act 1990. The key feature of the system of planning law operative in England and Wales is the requirement that planning permission is required for any "development" of land {s.57(1) TCPA 1990}. Development is defined as the carrying out of any building, engineering, mining or other operations in, on, over or under land, or the making of any material change in the use of any buildings or other land {s.55(1) TCPA 1990}.

Two essential elements in the planning control process are usefully distinguished. The first is the formulation of development plans by planning authorities to designate the potential uses of particular areas of land. The second is the actual determination of individual applications for development permission. Although both of these responsibilities are ultimately entrusted to the planning authorities, subject to ministerial guidance and supervision, the NRA has a consultative input into both processes.

In relation to the formulation of development plans, local planning authorities are under a duty to consult various government departments and public bodies (see Art.10 Town and Country Planning (Development Plan) Regulations 1991, S.I. 1991 No.2794). This allows bodies such as the NRA to comment upon the implications of a draft development plan in respect of any function exercised by the NRA.

With regard to individual applications for development consent, local planning authorities are required to publicise applications in a specified manner depending upon the nature of the application {Art.12B GDO 1988}. This allows any member of the public to make representations about the proposed development which the planning authority is bound to take into account in making its determination of the planning application {Art.22 GDO 1988}. In some instances, however, the requirement upon planning authorities to make planning information available becomes a *duty* to consult certain bodies termed "statutory consultees". Hence, for example, a planning authority will be bound to consult the NRA in relation to a development which involves the carrying out of works or operations in the bed or on the banks of a river or stream. Although the planning authority is not bound to follow the recommendations of a statutory consultee in respect of the eventual planning determination, it is bound to take recommendations into account in making a determination {Art.18 GDO 1988}.

Although formerly local planning authorities were only obliged to "have regard to" the appropriate development plan in determining particular applications for planning permission, this duty has been strengthened by the introduction of a more plan-led approach. Whilst a planning authority continues to be obliged to have regard to the provisions of the appropriate development plan, and to any other material considerations {s.70(2) TCPA 1990}, it is now further provided that determinations are to be made *in accordance with the plan* unless material considerations indicate otherwise {s.54A TCPA 1990}.

Development plans will take account of environmental implications as material considerations and specific requirements may be imposed in relation to groundwater protection. Hence it has been stated in official guidance:

"particular attention should be paid to the protection of groundwater resources which are susceptible to a wide range of threats arising from land use policies. Once groundwater has been contaminated, it is difficult if not impossible to rehabilitate it. Changes in land use may also affect the availability of groundwater resources by restricting recharge or diverting flows. The NRA has commissioned the preparation of a series of maps to identify those areas of particular concern. It is planned that these will become available over the next three years. They should be taken into account in drawing up development plans" (Department of the Environment, Planning Policy Guidance Note 12, *Development Plans and Regional Planning Guidance* (1992) para.6.19).

In respect of the duty upon a local planning authority to have regard to "other material considerations", this rather open ended duty will certainly require the environmental consequences of granting development to be taken into account. It will also require the authority to take into consideration any Ministerial guidance which is given in relation to the particular kind of development for which consent is sought. Accordingly, the authority will be under a duty to apply national planning policies as set out in Circulars and Planning Policy Guidance notes.

Whilst the main object of this section has been, primarily, to outline the role of the NRA in the planning process, it is also pertinent to note that departures from the general scheme of planning law operate in the context of highway planning control. Under the Highways Act 1980 central and local government authorities have extensive powers to provide new highways (see Section 3.2.3 below on highways generally). Broadly, the Secretary of State for Transport or the Secretary of State for Wales is empowered to designate land as the site for a trunk road or "special road" (see Section 3.2.3), and to acquire the required land either compulsorily or by agreement. Procedures are established requiring publicity and for persons affected by a designation order to object, and for their objections to be heard at a local inquiry conducted on behalf of the Secretary of State in a similar manner to a local planning inquiry. The Secretary of State is bound to consider the report of an inquiry before deciding whether to confirm the designation order (generally see Schedule 1 HA 1980).

Proposals for highway developments other than trunk or special roads are the responsibility of local highway authorities (see Section 3.2.3). Applications for planning permission in respect of developments of this kind of road are determined by the local planning authority, though provision is made for consultation with the local highway authority and the Secretary of State for Transport who may, under certain circumstances, issue directions to a local planning authority in respect of planning applications concerning highways {Arts.15 and 18 GDO 1988}. Specific kinds of material consideration are likely to arise in respect of highway planning applications which will need to be taken into account by the local planning authority in making a determination of this kind (see, in particular, Department of the Environment, Planning

Policy Guidance Note 13, *Highways Considerations in Development Control* (1988); and on matters of highway design see Department of Transport, *Design Manual for Roads and Bridges Vol. 10*).

Environmental Assessment

Of special significance in relation to highway and other major kinds of development is the provision for the environmental assessment of those projects which have a "significant effect" upon the environment. The origin of this requirement lies with the European Community directive on the assessment of the effects of certain public and private projects on the environment, the "Environmental Assessment Directive" ({85/337/EEC} and see Section 3.3 on Community directives generally). National legislation has been enacted in the United Kingdom to give effect to the requirements of this provision by subjecting certain kinds of development project to environmental assessment.

Accordingly, in relation to projects which fall within the scope of the directive, a prospective developer will be required to prepare an environmental assessment and the local planning authority, or any other body making a determination for which an environmental assessment is required, must give consideration to the environmental information supplied before development consent is granted. Specifically, the information to be provided must detail the effects of the proposed project upon fauna, flora, soil and water, including an estimate of the type and quantity of any expected residues and emissions resulting from the operation of the project (see Department of the Environment, *Circular 15/88*, and Department of the Environment, *Environmental Assessment A Guide to the Procedures* (1989).

Generally, environmental assessment is provided for under the Town and Country Planning (Assessment of Environmental Effects) Regulations 1988 {SI 1988 No.1199}. However, in relation to highways a special statutory provision has been enacted {s.105A(1) HA 1980, introduced under the Highways (Assessment of Environmental Effects) Regulations 1988 SI 1988 No.1241} which requires the Secretary of State, when considering the construction of a new highway or the improvement of an existing highway to determine, before publication of details of the project, whether it is likely to have a significant effect upon the environment. If this is the case, the requirements for environmental assessment must be complied with. By either legal mechanism, the environmental effects of a highway, or other development project, which is likely to have significant effects upon the environment, including the aquatic environment, are to be taken into account in the determination of the application for development consent. The Department of Transport *Design Manual for Roads and Bridges Vol. 11* deals specifically with Environmental Assessment.

Groundwater protection and planning

A particularly important aspect of the NRA's role in the planning process is indicated by their recent report, *Policy and Practice for the Protection of Groundwater* (1992). This report notes the NRA's general duty to monitor and protect the quality of groundwater and to conserve its use as a water resource {ss.84 and 19 WRA 1991}. The report also sets out a consistent national policy and a framework for decision making to which the NRA will adhere, for example, in response to consultation on planning applications.

The key principles of the policy for groundwater are, first, to protect the resources by relating activities to the vulnerability of the groundwater at particular locations as this is determined by the natural geological and other characteristics of the site. A second principle is to protect individual abstractions by defining source protection zones around each abstraction point where sources are used for public supply, private potable supply and for commercial food and drink production.

The application of these principles for groundwater protection leads to an explicit policy with regard to particular kinds of activity which may pose a threat to groundwater quality. Hence, to take a pertinent example, the NRA will seek to ensure that the line of any new major communication route avoids areas located immediately adjacent to a groundwater source

(Zone 1). Where new major roads are located within areas more remote from a source (Zone 2) the NRA will discourage the discharge of roadside drainage to underground water via soakaway systems because of the risk of spillage after accidents (*Policy and Practice for the Protection of Groundwater* (1992) p.47).

The NRA has stated its intention to use its role as a statutory consultee under planning law to influence determinations of those development permissions which may have an impact upon the yield or quality of groundwater. In addition, it is the express policy of the NRA to use its power to impose prohibitions upon particular discharges ({s.86 WRA 1991}, and see Section 3.4.1) where surface water runoff needs to be controlled in areas where groundwater is at risk. Discharges to groundwater will become subject to standard conditions, such as the installation of petrol/oil interception where this is applicable, and the NRA may require hydrological assessment to be carried out to identify the potential impact of the discharge on water resources (*Policy and Practice for the Protection of Groundwater* (1992) p.38).

The Water Resources Act 1991 provides for the designation of statutory water protection zones, in which activities which have an adverse effect upon controlled waters, including groundwater, may be made the subject of special legal prohibitions or restrictions imposed by the Secretary of State {s.93 WRA 1991}. However, it is not the present intention of the NRA to seek statutory designation of such zones, since it is thought that the combination of an active planning consultation role and effective control of particular discharges will be sufficient to counter the threat to groundwater. Despite this, the ultimate objective will be to ensure that the water quality objectives that are set for groundwater are achieved and maintained (see Section 3.3.2 below), and consequently it might become necessary in the future to reconsider the possibility of designating statutory water protection zones if other mechanisms for control prove to be inadequate.

3.2.2 River Purification Authorities in Scotland

The legal status of River Purification Authorities

Whilst overall executive responsibility for water in Scotland rests with the Secretary of State for Scotland, regulatory responsibilities for water pollution control are entrusted to the seven river purification boards and the three island councils (Orkney, Shetland and Western Isles). These bodies are collectively termed the "river purification authorities". Originally, river purification boards were established, with areas delineated by reference to water systems rather than local government boundaries, under the Rivers (Prevention of Pollution) (Scotland) Act 1951. This Act has been amended in various respects by the Rivers (Prevention of Pollution) (Scotland) Act 1965 and subsequent legislation (see River Purification Board Areas (Scotland) Order 1975, SI 1975 No.231). Corresponding functions were established for island councils under the Local Government (Scotland) Act 1973. Powers of the river purification authorities in relation to water pollution are provided for under Part II of the Control of Pollution Act 1974, though this has been substantially amended by measures under Schedule 23 to the Water Act 1989.

Under the Rivers (Prevention of Pollution) (Scotland) Act 1951, the Secretary of State is given the general duty of promoting the cleanliness of rivers, other inland waters and tidal waters of Scotland {s.1(1) R(PP)(S)A 1951}. More specifically, he is placed under a duty to establish river purification boards {s.135 Local Government (Scotland) Act 1973}. Originally, the membership of each board was to consist, in one third, of members appointed by the regional councils wholly or partly within the area. Another third was to consist of members appointed by the district councils, and the remaining third was to consist of members appointed by the Secretary of State after consultation with appropriate bodies with interests in agriculture, fisheries, industry and other relevant interests (see River Purification Boards (Establishment) Variation (Scotland) Order 1989, S.I. 1989 No.59). However, variation of these proportions has recently been provided for to allow the appointment of one quarter regional council appointees, one quarter district council appointees, and the remainder are to be appointees of the Secretary of State (see Schedule 10 para.6 Natural Heritage (Scotland) Act 1991).

Notably, the Scottish river purification authorities are more specialised bodies than the NRA in England and Wales, in that they do not possess a general range of functions relating to the aquatic environment. Generally, it is stated to be the duty of the river purification authorities to promote the cleanliness of the rivers in their areas, to conserve so far as practicable the water resources of their areas and to exercise for those purposes the functions conferred on them by the Rivers (Prevention of Pollution) (Scotland) Act 1951 {s.17(1) R(PP)(S)A 1951}. More specifically, the powers and duties of the river purification authorities are limited to matters of water quality monitoring and pollution control, since separate bodies have distinct responsibilities relating to drainage, fisheries and other aspects of water management. However, recent changes in the law have enabled river purification boards to control water abstraction under certain circumstances {Part II Natural Heritage (Scotland) Act 1991}.

Although river purification authorities are not subject to an explicit environmental duty of the kind imposed upon the NRA in England and Wales (see Section 3.2.1 above), the authorities are subject to the general duty imposed upon public bodies in Scotland with respect to the conservation of natural beauty. This requires that every Minister, government department and public body must have regard to the desirability of conserving the natural beauty and amenity of the countryside {ss.66 and s.78(1) Countryside (Scotland) Act 1967}. In relation to this duty, a Code of Practice for the river purification authorities is in the process of being prepared, dealing with environmental duties arising in relation to discharge consenting and other matters relating to the protection of the aquatic environment (*Draft Code of Practice on Conservation, Access and Recreation for Water, Sewerage and River Purification Authorities*, (1992) Scottish Office Environment Department).

Planning control and environmental assessment

Town and country planning law in Scotland generally follows the regime in England and Wales, described in Section 3.2.1, though with some notable points of difference. Significantly, overall responsibility for planning matters rests with the Secretary of State for Scotland, who provides guidance on matters of general planning policy ensuring consistent administration of a body of law based primarily upon the Town and Country Planning (Scotland) Act 1972 as amended. More specific matters of plan formulation and the determination of applications for development consent are entrusted to district, regional and general planning authorities, with district councils, in the capacity of "district planning authorities", having primary responsibility for local plans and development control.

Whilst the general procedures in relation to determinations of applications for planning permission parallel those in England and Wales, it is pertinent here to note the status of the river purification authorities in planning law. In relation to the formulation of a local plan, a planning authority must consult with all local authorities in the area concerned and also "such other authorities and bodies as it thinks appropriate" {Reg.4(2)(a) Town and Country Planning (Structure and Local Plans) (Scotland) Regulations 1983, SI 1983 No.1590}. Implicitly, therefore, river purification authorities will be allowed the opportunity to comment upon development plans prior to their approval. Also, river purification authorities will be statutory consultees in relation to particular development proposals which may have an adverse effect upon the aquatic environment {Art.13(2) Town and Country Planning (General Development) (Scotland) Order 1981, SI 1981 No.830}. In general terms, therefore, the position of river purification authorities in planning law compares with that of the NRA in England and Wales.

Specifically in the planning of roads, nothing in the Roads (Scotland) Act 1984 authorises the carrying out of any development for which planning permission is required under the Town and Country Planning (Scotland) Act 1972 {s.123 RSA 1984}. An exception to this arises where a roads authority determines or redetermines the extent of the right of public passage on a road, where change of use and consequent works are deemed to be given planning permission under the 1972 Act {s.152(2) and (4) RSA 1984}. Normally, however, the construction of roads will be subject to the general planning process either as provided for under the Roads (Scotland) Act 1984 or the Town and Country Planning (Scotland) Act 1972.

In accordance with the European Community Environmental Assessment Directive {85/337/EEC}, there is a specific obligation upon roads authorities (see Section 3.2.4) to consider the impact of road developments upon the environment and, where necessary, to produce an environmental impact assessment, identifying, describing and assessing the direct and indirect effects of the project on the environment {s.20A RSA 1984, substituted by the Environmental Assessment (Scotland) Regulations 1988, SI 1988 No.1221}. In relation to the Secretary of State for Scotland the requirement of environmental assessment is imposed under roads legislation, and in respect of local roads authorities the requirement is imposed under the planning process.

3.2.3 Highway responsibilities in England and Wales

Highway law generally

In legal terms, in England and Wales, a "highway" is defined as a way over which the public has a right to pass freely, without hindrance at all times of the year. Highways may be created at common law by the owner of the land dedicating the right of passage to the public and public acceptance of that right, and long use by the public may be evidence of dedication. However, of greatest practical importance is the facility for highways to be created by "highway authorities" in England and Wales using statutory powers provided for under the Highways Act 1980.

Under the Highways Act 1980, except where the context otherwise requires, "highway" means the whole or a part of a highway other than a ferry or waterway, and where a highway passes over a bridge or through a tunnel, that bridge or tunnel is taken to be a part of the highway {s.328(1) and (2) HA 1980}. Highways may be distinguished in law according to the extent of public use which is permissible upon them, and by the express designation of certain roads as trunk roads, classified roads or special roads. On some highways public use is limited to a particular class of traffic, for example, with footpaths and bridleways. In respect of the particular highway designations, special statutory provision has been made. Hence, trunk roads were formerly provided for under the Trunk Roads Act 1936 (now see {s.10 HA 1980}), classified roads were provided for under the Local Government Acts 1966 and 1974 (now see {s.12 HA 1980}) and special roads, reserved for special classes of traffic, were provided for under the Special Roads Act 1949 (now see {s.16 HA 1980}, and the Motorway Traffic Regulations 1982, SI 1982 No.1163).

Highway Authorities

Creation, improvement and maintenance of highways are principally the concern of the "highway authorities", meaning either the Ministers with transport responsibilities or local highway authorities. In relation to trunk roads and special roads, the highway authority is the Secretary of State for Transport or the Secretary of State for Wales. The highway authority in respect of most roads other than trunk roads is the county or metropolitan district council or the London borough council or the Common Council of the City of London {s.1 HA 1980}.

A local highway authority may construct new highways {s.24(2) HA 1980}. Alternatively, a private carriage or occupation road may be dedicated by the owner and become a highway maintainable at the public expense by an agreement with a highway authority on such terms as may be agreed {s.38(3)(A) HA 1980}. In addition, an owner of land may agree with a highway authority to dedicate as a highway, to be maintainable at the public expense, a way to be constructed by him on such terms as may be agreed. Such an agreement will provide, amongst other things, that the way shall be properly sewered, levelled, paved, metalled, flagged, channelled, drained and joined with an existing road to which it abuts and generally constructed in accordance with agreed specifications.

The general position is that a highway which is maintainable at public expense, together with the materials and scrapings of it, vests in the appropriate highway authority {s.263 HA 1980}. However, the effect of vesting does not confer upon a highway authority any rights in relation to mines and minerals under the highway vested in that authority {s.335 HA 1980}. The drains belonging to a road vested in a local highway authority are also vested in that authority, and where any other drain or sewer was used for the drainage of the road at the time when it first became maintainable at public expense, the authority will continue to have the right to use that drain or sewer for that purpose {s.264 HA 1980}.

Duties of Highway Authorities

The precise extent of the highway for the purpose of the duty to maintain is a question of fact, but it is evident that it will extend to the provision of drains which are part of the highway. For the purpose of draining a highway, or of otherwise preventing surface water from flowing onto it, the highway authority may conduct certain operations. These allow it to construct or lay in the highway, or in land adjoining or lying near it, such drains as it considers necessary; erect barriers in the highway or in such land to divert surface water into or through any existing drain; and scour, cleanse and keep open all drains situated in the highway or in such land ({s.100(1) HA 1980}, and see Section 3.4.3). For these purposes the expression "drain" is defined to include a ditch, gutter, watercourse, soak-away, bridge, culvert, tunnel and pipe {s.100(9) HA 1980}.

Water which has been drained and diverted from a highway may then be discharged into any inland waters or tidal waters, but the consent of the NRA, or other drainage body within the meaning of the Land Drainage Act 1991, or navigation authority will be required where the exercise of these powers interferes with a watercourse vested in such an authority {s.339 HA 1980}. Moreover, the powers to undertake highway drainage operations are without prejudice to any enactment, the purpose of which is, to protect water against pollution ({s.100(8) HA 1980} and see Section 3.4).

The Water Industry Act 1991 enables highway authorities and sewerage undertakers to enter into reciprocal agreements, whereby a highway authority may use any public sewer vested in the sewerage undertaker for the conveyance of surface water from roads repairable by them, and a sewerage undertaker may use any drain or sewer vested in the highway authority for the purpose of conveying surface water from premises or streets {s.115 WIA 1991}. Neither an authority nor a sewerage undertaker may unreasonably refuse to enter into an agreement of this kind, and neither the authority nor the undertaker may insist unreasonably upon terms unacceptable to the other party. Although no obvious legal solution is readily available, the overlap of responsibilities which may arise in respect of the shared use of public sewers or highway drains may raise potential legal difficulties as to the respective responsibilities of the two bodies concerned in the event of the use of a shared facility giving rise to pollution problems.

A private individual has no right to connect his drains or sewers to a highway drain and is liable for any nuisance arising as a result of an unauthorised connection {*Wincanton RDC v. Parsons* [1905] 2 *King's Bench Law Reports* 34}. Moreover, it is an offence to alter, obstruct or interfere with a drain which has been constructed or laid by a highway authority without the authority's consent, and the authority may also recover from such a person its expenses of repair or reinstatement necessitated by his action {s.100(4) HA 1980}.

As previously indicated, a potential problem with respect to water pollution originating from highways in England and Wales arises from activities which are undertaken by a highway authority for purposes of highway maintenance. Of particular concern in the present context are operations such as the application of substances for de-icing and herbicidal purposes. The legal justification for these activities originates in duties imposed under the Highways Act 1980, but the use of particular chemical substances may also be regulated under separate legislation.

A highway authority is placed under a general duty to maintain a highway which is maintainable at public expense {s.41 HA 1980} and, more specifically, to remove obstructions which may arise on a highway by the accumulation of snow or from any other cause {s.150(1) HA 1980}. The precise extent of this duty to maintain highways is not made explicit under the

1980 Act, but some guidance is provided by case-law in which it has been held that a highway authority's duty is reasonably to maintain and repair a highway so that it is free of danger to all persons who use the highway in the way normally expected of them. In performing that duty, the authority has to provide not merely for the model motorist, who always drives with reasonable care, but for the normal run of drivers including those who make mistakes of a kind which experience and common sense has taught are likely to occur. Hence, the test to apply is whether the condition of the road is foreseeably dangerous to vehicles being driven in a normal, rather than exemplary, manner {*Rider v. Rider* [1973] 1 *All England Law Reports* 294}. Notably, the duty to maintain a highway is wider in scope than keeping in repair the surface of the road, and may include the removal of snow or ice or taking such protective measures as would render highways safe for road users in bad weather conditions {*Haydon v. Kent County Council* [1978] 2 *All England Law Reports* 97}.

In practice, an important maintenance duty of highway authorities of relevance to the present study is the duty to keep the roads, and roadside areas, for which it is responsible clear of vegetation. Alongside maintenance and safety considerations, in respect of certain injurious weeds, there is a statutory duty on the occupier of any land on which they are growing to prevent them from spreading to adjacent land {s.3(1) Weeds Act 1959}. It is apparent, therefore, that highway authorities have a duty to ensure that the lands on which roads are situated do not become a nuisance to neighbouring land through the spread of certain injurious weeds.

In exercising the duty to keep roads clear from vegetation, however, highway authorities may become subject to separate legislation governing the application of herbicidal preparations. The application of herbicides of various kinds will be subject to a body of specialised legislation under Part III of the Food and Environment Protection Act 1985 and the Control of Pesticides Regulations 1986 {S.I. 1986 No.1510}, providing for Ministerial approval of a particular pesticide and consent for its use in specified circumstances. Specifically, persons making use of pesticides must have received adequate instruction and guidance in the safe, efficient and humane use of pesticides and be competent for the duties they are required to perform. Users of pesticides must also take all reasonable precautions to protect the health of human beings, creatures and plants, and to safeguard the environment. Additionally, the use of herbicides is likely to be subject to the Control of Substances Hazardous to Health Regulations 1988 {S.I. 1988 No.1657}, which regulate the exposure of employees to substances in an approved list and requires employers to ensure that exposure of employees is prevented or adequately controlled. General guidance as to the safeguards which are needed in the use of herbicides is provided in literature issued by the Department of the Environment (see *Weed Control and Environmental Protection* and *Guidance for Control of Weeds on Non-agricultural Land* (1992)).

3.2.4 Roads responsibilities in Scotland

Roads law generally

In Scotland a "road" is defined as any way, other than a waterway, over which there is a public right of passage, by whatever means and whether subject to toll or not, and includes the road's verge, and any bridge, whether permanent or temporary, over which, or tunnel through which, the road passes. Any reference to a "road" includes a part thereof {s.151(1) RSA 1984, as amended}. Accordingly, a road will generally include the whole area dedicated to public passage, extending from fence to fence or from building line to building line {*John Munro Beattie v. Scott* [1990] *Scottish Criminal Case Reports* 435}.

Roads may be either "public" or "private" depending upon whether or not they are maintained by a roads authority. Whilst various minor categories of road are provided for under the 1984 Act, including "footways", "footpaths" and "cycle tracks", those roads for general vehicular traffic are termed "carriageways" {s.151(2) RSA 1984}.

With respect to ownership, public roads which are the responsibility of the local roads authority vest in that authority for the purpose of exercising its functions, and similarly those public roads maintainable by the Secretary of State are vested in him {s.1(9) and s.2(4) RSA 1984}. However, the ground on which the road is situated, termed the "solum", does not belong to the local roads authority or the Secretary of State unless it has actually been purchased, and usually vests in the owners of the adjoining land {s.115(1) RSA 1984}.

Special roads are provided by a "special road authority" and constructed, or appropriated by that authority along a route prescribed by a scheme for the use of classes of traffic specified by the scheme {ss.7 and 8 and Schedule 3 RSA 1984}. A special road scheme may be made by the Secretary of State or by a local roads authority in accordance with regulations and subject to his confirmation ({s.7(6) RSA 1984} and see {Environmental Assessment (Scotland) Regulations 1988, SI 1988 No.1221}).

A roads authority may acquire land, or an interest in land, where this is necessary for the purposes of its functions {ss.104 and 110(2) RSA 1984}. Land may be acquired compulsorily or by agreement {s.103 RSA 1984}, but a local roads authority will require the authorisation of the Secretary of State before exercising any power of compulsory acquisition {s.110(1) RSA 1984}. Further powers allow a roads authority to acquire land for the purpose of mitigating any adverse effect of a road on the environment {s.106 RSA 1984}. Owners or occupiers of land adjoining a road may be required by a roads authority to prevent water or other matter flowing or percolating onto a road from the adjoining land and may be required to carry out works to prevent this {s.99 RSA 1984}.

Roads Authorities

The principal enactment concerning roads in Scotland is the Roads (Scotland) Act 1984, consolidating and codifying a large number of previous Acts. The 1984 Act, which was brought fully into force on 8 January 1991, includes a range of provisions relating to the present study. Broadly, the following matters are provided for: the powers and duties of roads authorities, consisting of either the Secretary of State for Scotland or regional or island councils; public, trunk, special and private roads; the creation, improvement and maintenance of roads; the control of work on roads; the diversion of waters when constructing or improving roads; the acquisition, transfer and vesting of land and for the rights of statutory undertakers.

Responsibilities for roads in Scotland are divided between the Secretary of State for Scotland and local roads authorities. The Roads (Scotland) Act 1984 assigns executive functions relating to the overall roads network to the Secretary of State along with operational powers and duties in respect of the planning, management and maintenance of trunk roads, special roads and certain other roads for which he has responsibility. Otherwise, responsibility for other public roads lies with the nine regional or three island councils, which act as the "local roads authorities" within their respective areas. The powers and duties of local roads authorities include the planning, construction, maintenance and management of those roads for which they are responsible.

The Secretary of State for Scotland has ultimate responsibility for Government policy and administration in relation to roads in Scotland. It is his duty to classify roads, keep under review the national system for through traffic routes and designate roads as trunk roads {s.5(2) RSA 1984}. He has extensive powers to enact regulations and make schemes and orders concerning special roads and other matters, and certain actions taken by local authorities are required to be confirmed, authorised or consented by him before they come into effect. In relation to any matter on which he is authorised to act, or any determination which he is authorised to make under the 1984 Act, he may hold a local inquiry and in relation to certain matters he is required to hold one {s.139(1) RSA 1984}.

Alongside his overall responsibility for Government policy and administration in respect of roads, the Secretary of State exercises a separate operational function as the roads authority in respect of certain categories of road. Specifically, as a roads authority he is required to manage and maintain trunk roads, special roads and other roads constructed by him which have not been listed as maintainable by a local roads authority {s.2(1) RSA 1984}. The Secretary of State may construct new trunk roads or other roads with the consent of the Treasury {s.19

RSA 1984}, and special roads if authorised to do so by a special roads scheme {s.7(3) RSA 1984}. He has the power to reconstruct, alter, widen, improve or renew the roads which he manages and maintains, and to determine how the public right of passage over these roads may be exercised {s.2(1) RSA 1984}. "Improvement" for these purposes encompasses keeping the road in a proper state of repair and doing anything for the benefit of road users, including the improvement of the amenity of the road and adjacent land {s.151(1) RSA 1984}.

In practice, the powers and duties given to the Secretary of State are carried out on his behalf by the Roads Directorate within the Industry Department of the Scottish Office. The technical and engineering divisions of the Roads Directorate undertake the preparation and execution of schemes for new roads along with maintenance, improvement and other matters, though the design and construction of particular schemes is usually undertaken by independent consulting engineers or local roads authorities under the supervision of the Directorate.

Routine maintenance of roads vested in the Secretary of State is usually carried out on his behalf by local roads authorities in accordance with agreements and codes of practice {s.4(1) RSA 1984}. The "cleansing" of roads by the removal of litter or refuse is usually undertaken by district or island councils in accordance with similar arrangements ({s.4(1) RSA 1984}, and see {ss.86 and 96 Environmental Protection Act 1990}). Where maintenance or cleansing arrangements are entered into, however, the local roads authority or the district or island council acts only as an agent for the Secretary of State who remains the roads authority for the road concerned {s.4(4) RSA 1984}.

The overall responsibility of the Secretary of State for trunk and special roads will encompass duties with respect to the de-icing treatment of roads and herbicidal treatment of roadside vegetation. The nature of these responsibilities, and their water pollution control implications, are described below.

Provision is made for the Secretary of State and a local authority to enter into agreements as to the use of sewers and drains for the conveyance of water from the surface of a trunk road. Neither the Secretary of State nor a local authority may unreasonably refuse to enter into such an agreement {s.7 Sewerage (Scotland) Act 1968}. Because local roads authorities and sewerage authorities are within the regional council structure in Scotland, there is no need for explicit provision for reciprocal agreements between highway authorities and sewerage undertakers as exist in England and Wales (see discussion of {s.115 WIA 1991} in Section 3.2.3).

The regional or island councils act as the local roads authority for those public roads within their areas which are not the responsibility of the Secretary of State. A local roads authority is required to maintain a list of roads which it manages and maintains within its area {s.1(1) RSA 1984}. Where a road is entered into this list it vests in the local roads authority for the purpose of its functions as a roads authority, but the authority will not become the owner of the ground upon which the road is situated unless this is expressly conveyed to it {s.1(9) RSA 1984}.

A local roads authority is empowered to construct any new road which it considers is required, other than a special road, unless it is empowered to do so under a special road scheme {s.20(1) and (3) RSA 1984}. A local roads authority may reconstruct, alter, widen, improve or renew any existing road which it manages and maintains {s.1(1) RSA 1984}.

Duties of Roads Authorities

As has been noted, both the Secretary of State and local roads authorities have statutory duties to manage and maintain all the roads for which they are responsible, and a general power to reconstruct, alter, widen, improve or renew these roads {ss.1(1) and 2(1) RSA 1984}. Accordingly, a failure to maintain and repair a road on its list will make a road authority liable if injury or loss is suffered due to a roads authority's failure to keep the road in a safe condition {*Threshie v. Annan Magistrates* (1845) 8 *Session Cases 2nd Series (Dunlop)* 276}.

A particular application of the general duty of roads authorities to manage and maintain roads, of special relevance to this study, is the duty of a roads authority to execute works for the purpose of draining a road or otherwise preventing surface water from flowing onto it {s.31 RSA 1984}. Roads authorities are also authorised to contribute towards any expenses incurred in the execution of maintenance of works under the Land Drainage (Scotland) Act 1958 or flood prevention operations under the Flood Prevention (Scotland) Act 1961 {s.32 RSA 1984}.

In order to drain a public road, or proposed public road, and to prevent surface water from flowing onto it, a roads authority may construct or lay drains in the road or land adjoining it; erect or maintain barriers to divert surface water into or through any existing drain; scour cleanse and keep open drains; and drain surface water into any natural or artificial inland waters or tidal waters {s.31(1) RSA 1984}. For these purposes a "drain" is stated to include a ditch, gutter, watercourse, bridge, culvert, pipe or holding pond and any pumping machinery associated with any of those things {s.31(6) RSA 1984}.

Before carrying out drainage works, the roads authority must serve written notice of its intention to carry out the work on the owners and occupiers of land affected by the works, who may object to the works {s.31(2) RSA 1984}. If an objection is sustained then, if the Secretary of State is the roads authority, he must consider the objection before coming to a decision as to whether or not to carry out the works. Alternatively, if the works are being carried out by a local roads authority, the matter must be referred to the Secretary of State, who may give his consent to the works and, if necessary, subject them to conditions {s.31(4) RSA 1984}.

A person who suffers damage in the execution of works by a roads authority in accordance with its powers, is entitled to compensation from the authority {s.116(1) RSA 1984}. The precise meaning of "damage" in this context is not defined, but any dispute as to the amount of compensation payable is to be determined in accordance with the Land Compensation (Scotland) Act 1963 {s.117(1) RSA 1984}. Whilst the statutory right to compensation will only arise in relation to damage sustained by reason of the drainage works, this right is stated to be without prejudice to other rights to compensation. That is, there may be, for example, situations where compensation is payable under the general civil law which fall outside the right to statutory compensation in respect of drainage operations (see Section 3.5.3 on civil law).

A roads authority will be legally liable for any injury or loss resulting from its failure to keep a road in a reasonably safe condition {*Virtue v. Alloa Police Commissioners* (1873) 1 *Session Cases 4th Series (Rettie)* 285}. In particular, this liability may arise, for example, by a failure to take the necessary action to protect, maintain or repair it {*Laing v. Paull and Williamsons* (1912) *Session Cases* 196}, or in respect of a defective sewer grating {*Rush v. Glasgow Corporation* (1947) *Session Cases* 580}.

In addition to the general maintenance responsibilities of roads authorities, an explicit statutory responsibility is imposed in relation to the clearance of snow and ice. This is that roads authorities are bound to take reasonable steps to prevent snow and ice endangering the safe passage of pedestrians and vehicles over public roads {s.34 RSA 1984}. Notably, however, this is not an absolute duty, in the sense of requiring all roads to be free from all snow and ice at all times, and an authority will have a discretion to decide upon priorities in clearing roads according to their respective importance ({*Cameron v. Inverness County Council* (1935) *Session Cases* 493} and {*Gordon v. Inverness Town Council* (1957) *Scots Law Times* 48}).

The counterpart of the duty upon roads authorities to take reasonable steps to clear roads of snow and ice is that a roads authority may be held liable for failing to exercise the duty effectively, for example, by allowing the surface of a road to become slippery or dangerous {*Cairnie v. Secretary of State for Scotland* (1966) *Scots Law Times* 57}. More generally, it has been held that a road surface will be dangerous if, due to adverse weather conditions or otherwise, it is likely to be a hazard to road users. In ascertaining whether a hazard exists it must be anticipated that normal traffic will occasionally include vehicles being driven with less than all due skill and care {*Rae v. Dunbarton County Council* (1973) *Scots Law Times* 23}.

In relation to the present study, therefore, it is apparent that the application of de-icing preparations to road surfaces will be undertaken as a legal obligation. Nevertheless, the legal duty to apply potentially polluting substances to roads does not, by itself, provide any explicit defence to offences which may arise in respect of the pollution of water.

As in relation to England and Wales, a roads authority will be under a duty to keep the roads, and roadside areas, for which it is responsible clear of vegetation. The same statutory duty is imposed upon an occupier of any land on which certain weeds are growing to prevent them from spreading to adjacent land {s.3(1) Weeds Act 1959}. Similarly, the Control of Pesticides Regulations 1986 {SI 1986 No.1510}, enacted under Part III of the Food and Environment Protection Act 1985 will be applicable to the use of a pesticide on a road (see Section 3.2.3). Also under the Control of Pollution Act 1974, it is an offence to cause pollution by pesticide/herbicide spraying. Likewise the Control of Substances Hazardous to Health Regulations 1988 {SI 1988 No.1657} under the Health and Safety at Work etc. Act 1974 will impose further safety requirements upon the use of controlled substances (see Section 3.2.3).

3.2.5 The transport of polluting substances by road

A significant concern in respect of pollution originating from highways is that of pollutants which may enter highway drainage systems as a consequence of spillages arising from traffic accidents. In respect of this concern, legal controls exist both in relation to types of vehicles that may be used to carry hazardous substances and in relation to procedures that must be followed in carrying these substances by road. Whilst, primarily, the legislation is intended to meet health and safety requirements, it is apparent that the measures enacted are of equal importance in respect of the environmental protection in providing important safeguards against the escape of many potentially polluting substances onto roads and into watercourses or groundwater, either directly or as a result of clean-up operations.

Whilst general requirements for the roadworthiness of vehicles are provided for under the Road Traffic Act 1988 (as amended by the Road Traffic Act 1991), specific requirements relating to the road transport of hazardous substances are provided for under the Health and Safety at Work etc. Act 1974. This Act gives effect to various international obligations including the European Agreement concerning the International Carriage of Dangerous Goods by Road (1957, and subsequently revised). The 1974 Act allows for the enactment of particular regulations concerned with health and safety by statutory instrument {ss.15 and 82(3) HSWA 1974}. Specifically, regulations of this kind may be made for the purposes of prohibiting or regulating the transport of articles or substances of specified descriptions, and imposing requirements with respect to the manner and means of transporting them. These requirements may relate to the construction, testing, marking of containers and means of transport, and the packaging and labelling of articles or substances in connection with their transport {Schedule 3 para.3 HSWA 1974}.

Of greatest general relevance to the present discussion are the Road Traffic (Carriage of Dangerous Substances in Road Tankers and Tank Containers) Regulations 1992 {SI 1992 No.743}, but see also the Road Traffic (Carriage of Dangerous Substances in Packages etc.) Regulations 1992, {SI 1992 No.742}. The Regulations relating to road tankers and tank containers apply to the road transport of dangerous substances on the "Road Tanker Approved List" published by the Health and Safety Commission {under Reg.4(1)}. Duties are imposed upon the operators of road tankers, and tank containers used to transport these dangerous substances, to ensure that their vehicles are properly designed, of adequate strength and of good construction from sound and suitable material, and fit for the purpose for which they are used {Reg.6}. Moreover, they are to be properly maintained and subject to periodic examination and testing {Reg.7}.

The Regulations make it unlawful for specified dangerous substances to be transported by road {Reg.11}. Precautions must be taken against fire and explosion {Reg.13}, and road tankers carrying dangerous substances are to carry hazard warning panels {Reg.18}. An operator is not permitted to transport a dangerous substance unless he has sufficient information from the consignor to enable him to appreciate the risks involved {Reg.10}. The operator must also ensure that the driver of the vehicle has adequate instruction and training to understand the nature of the dangers to which the substance being carried may give rise, and the actions that should be taken in an emergency concerning it {Reg.26}.

Failure to adhere to the requirements imposed under the Road Traffic (Carriage of Dangerous Substances in Road Tankers and Tank Containers) Regulations 1992 is a criminal offence under the Health and Safety at Work etc. Act 1974 {s.33 HSWA 1974}, though it will be a defence to show that all reasonable precautions have been taken and all due diligence exercised to avoid committing the offence {Reg.29}. Breach of the Regulations may also give rise to civil liability {s.47 HSWA 1974}.

For Scotland, as for England and Wales, the Health and Safety at Work etc. Act 1974 allows for the enactment of regulations concerning the transport of hazardous substances by road. Accordingly, the Road Traffic (Carriage of Dangerous Substances in Road Tankers and Tank Containers) Regulations 1992 {SI 1992 No.743} and the Road Traffic (Carriage of Dangerous Substances in Packages etc). Regulations 1992 {SI 1992 No.742}, described above, will apply similarly in Scotland as in England and Wales.

3.3 THE EUROPEAN COMMUNITY AND NATIONAL WATER QUALITY LAW

3.3.1 European Community law in England and Wales

Whilst the legislation of England and Wales providing for the criminal offence of water pollution is described later (see Section 3.4), this largely amounts to a reactive approach to particular instances of water pollution by providing for punishment of offenders after the event. Clearly prevention is better than cure in relation to the protection of the aquatic environment, and over recent years an increasing need has been recognised to adopt a more proactive and strategic approach to problems of water quality. In legal terms this is to be accomplished, in part, through legislation which strives to realise and maintain water quality objectives relating to the categories of water use. This progression towards strategic regulation of water quality is a consequence of legal developments at both European and national levels.

At the European Community level, there have been enacted a large number of directives concerned with protection of the environment. As a matter of Community law the status of directives is provided for by the Treaty of Rome 1957, which states that "a Directive shall be binding, as to the result to be achieved upon each member state to which it is addressed, but shall leave to the national authorities the choice of form and methods" {Art.189 Treaty of Rome 1957}. That is to say, a member state is legally bound to give effect to the obligations arising under a directive within its territory, though some flexibility is provided for in the precise manner in which national legal systems and administration are used to accomplish this. Nevertheless, "competent authorities" must be designated within each member state, and in respect of water quality directives, in England and Wales, the competent authority for most purposes is the NRA. Accordingly, they will be subject to obligations originating from European law though, as will be seen, national laws have been enacted to secure effective implementation of many Community obligations relating to the aquatic environment (see Section 3.3.2).

From amongst the list of European Community water quality directives two measures are likely to be of greatest significance in relation to pollution from highways: the directive on pollution caused by certain dangerous substances discharged into the aquatic environment of the Community, known as "the Dangerous Substances Directive" {76/464/EEC}, and the directive on the protection of groundwater against pollution caused by certain dangerous substances, known as "the Groundwater Directive" {80/68/EEC}. The former will be particularly relevant to surface water pollution from highway runoff, whilst the latter is a counterpart which is important in relation to groundwater pollution originating from highways.

The Dangerous Substances Directive

The Dangerous Substances Directive seeks to establish a framework for the elimination or reduction of pollution by certain specified dangerous substances in inland, coastal and territorial waters. This is done by the classification of substances according to their harmfulness as List I or List II substances. List I encompasses the most toxic, persistent and bio-accumulatable substances and includes organo-halogens, organo-phosphorous and organo-tin compounds, mercury, cadmium and their compounds, and substances possessing carcinogenic properties. List II encompasses less harmful substances including metals such as zinc, copper, nickel, chromium, lead and their compounds, biocides and their derivatives, and substances such as cyanide and ammonia.

The fundamental obligation under the directive is that the member states of the Community are to take steps to eliminate pollution by List I substances and to reduce pollution by those in List II. Specifically, the definition of pollution which is applied for these purposes is "the discharge by man, directly or indirectly, of substances or energy into the aquatic environment, the results of which are such as to cause hazards to human health, harm to living resources and to aquatic ecosystems, damage to amenities or interference with other legitimate uses of water" {Art.1 para.2(e) DSD 1976}.

Discharges liable to contain List I and List II substances are subject to a requirement of prior authorisation by the competent authority of the member state concerned {Art.3 para.1 and Art.7 para.2 DSD 1976}. Member States are to establish pollution reduction programmes for List II substances based upon quality objectives and subject to implementation deadlines {Art.7 DSD 1976}.

Competent authorities within member states are to draw up an inventory of the discharges into their waters which may contain List I substances to which emission standards are applicable {Art.11 DSD 1976}. This information, along with details of authorisations for List I and List II substances, and the results of monitoring and information on pollution reduction programmes, are to be supplied to the Commission at its request {Art.13 DSD 1976}.

The Groundwater Directive

Originally, groundwater protection had been provided for under the Dangerous Substances Directive, but since 1980 it has been covered separately under the Groundwater Directive {80/68/EEC}. This directive strives to prevent the pollution of groundwater by substances identified under Lists I or II of the Annex to the Directive, which are similar to those under the Dangerous Substances Directive, and as far as possible to check or eliminate the consequences of pollution which has already occurred {Art.1(1) GD 1980}. For the purposes of the directive "groundwater" means all water which is below the surface of the ground in the saturated zone and in direct contact with the ground or subsoil, and "pollution" means the discharge by man, directly or indirectly, of substances or energy into the groundwater, the results of which are such as to endanger human health or water supplies, harm living resources and the aquatic ecosystem or interfere with other legitimate uses of water {Art.1(2) GD 1980}.

The key obligations of the directive are that member states are to take the necessary steps to prevent the introduction into groundwater of substances in List I, and to limit the introduction into groundwater of substances in List II so as to avoid pollution of this water by these substances {Art.3 GD 1980}. In order to comply with these obligations, member states are to prohibit all direct discharges of substances in List I, and to take appropriate measures to prevent any indirect discharges of List I substances owing to activities on or in the ground {Art.4(1) GD 1980}. "Direct discharge" means the introduction into groundwater of substances in Lists I or II without percolation through the ground or subsoil, whilst "indirect discharge" means the introduction into groundwater of substances in Lists I or II after percolation through the ground or subsoil {Art.1(2) GD 1980}.

Member states are to make all direct discharges of List II substances the subject of prior investigation of the hydrological and other conditions of the area concerned, so as to limit such discharges, and similarly to limit the disposal or tipping of those substances which might lead to indirect discharge. In the light of these prior investigations, the competent authorities in member states may grant an authorisation for the direct or indirect discharge or disposal of List II substances, provided that all the technical precautions for preventing groundwater pollution by these substances are observed {Art.5(1) GD 1980}.

Where a discharge is authorised by the competent authority in a member state, the authorisation is to specify a range of particular matters. The details to be specified include essential precautions, with particular attention being paid to the nature and concentration of the substances present in the effluents, the characteristics of the receiving environment, and the proximity of water catchment areas, in particular those for drinking, thermal and mineral water {Art.9 GD 1980}. The competent authorities are to monitor compliance with the conditions laid down in the authorisations and the effect of discharges on groundwater {Art.13 GD 1980}.

Overall, it may be noted that the directive is stated not to be applicable to discharges containing substances in Lists I or II in a quantity and concentration so small as to obviate any present or future danger of deterioration in the quality of receiving groundwater {Art.2 GD 1980}. Arguably, uncontaminated surface water runoff from highways should not be covered by the directive. However, it is stated that an authorisation for discharge to groundwater may only be granted where there is no *risk* of polluting the groundwater, and the conditions to which an authorisation is subject have to specify essential *precautions*. Accordingly, the *possibility* that, for example, an infiltration system designed to transmit surface water from a highway *might* receive polluting matter and channel it into groundwater, for example from an accidental spillage of some kind, will mean that this is a consideration to be taken into account.

3.3.2 National Water Quality Objectives in England and Wales

Water directives and national law

As has been explained, the Community water directives impose a duty upon each member state to give legal effect to the obligations arising under the directives by means of national legislation. Specifically, national laws are required to be enacted to ensure that water which falls within the scope of each particular directive meets the quality parameters stipulated by that directive. Likewise there is required to be a national body, termed the "competent authority", to oversee monitoring and law enforcement within the scope of each directive within each member state.

In relation to England and Wales the competent authority for the water directives is generally the NRA, though in certain respects, as under the Groundwater Directive, other bodies are involved as competent authorities in relation to waste on land. Although different water directives require a different mechanism for their implementation, in national law an important mechanism for transposition is through Chapter I of Part III of the Water Resources Act 1991, concerned with "Quality Objectives" {ss.82 to 84 WRA 1991}. Three essential features of the legal mechanism for giving effect to Community obligations in national law are to be discerned: the classification of water quality; the specification of water quality objectives; and the general duty to achieve and maintain objectives.

Classification of quality of waters

Although in England and Wales it has been a long-standing national administrative practice to classify waters according to their suitability for various purposes, a distinctive feature of the Water Act 1989, and now the Water Resources Act 1991, is the provision for *statutory* water classification systems. An enabling power allows the Secretary of State to prescribe a system of classifying the quality of controlled waters according to criteria specified in the regulations {s.82(1) WRA 1991}. These criteria are to consist of: (a) general requirements as to the purpose for which the waters to which the classification is applied are to be suitable; (b) specific requirements as to the substances that are to be present in or absent from the water and as to the concentrations of substances which are, or are to be, present in the water; and (c) specific requirements as to other characteristics of those waters {s.82(2) WRA 1991}.

The programme of enacting statutory water quality classification regulations has not yet encompassed all the different kinds of waters covered by Community directives, but some water quality classification regulations have been created. For example, the Surface Waters (Classification) Regulations 1989 {SI 1989 No.1148} prescribe a system for the classification of the quality of inland waters according to their suitability for supply, after treatment, as drinking water. Similarly, the Surface Waters (Dangerous Substances) (Classification) Regulations 1989 {SI 1989 No.2286} prescribe a system for the classification of "dangerous substances" in inland and coastal waters.

Water Quality Objectives

Following the enactment of statutory water classification systems, the second stage in the implementation process is the facility for applying particular quality classifications to individual waters. Accordingly, for the purpose of maintaining and improving the quality of controlled waters, the Secretary of State is empowered to serve a notice on the NRA specifying one or more of the prescribed water quality classifications and a date for compliance. This specification will serve to establish the water quality objectives for the waters concerned {s.83(1) WRA 1991}. The realisation of any specified water quality objective for a particular water is the satisfaction by that water of the appropriate classification requirements on, and at all times after, the stipulated date {s.83(2) WRA 1991}.

At the level of policy formulation the NRA has been active in drawing up proposals and recommendations for water quality objectives. The recently published *Proposals for Statutory Water Quality Objectives* (National Rivers Authority, 1992), which is presently under consideration by the Department of the Environment (*River Quality* 1992), provides for a scheme of water quality targets based upon six "use" classes: fisheries ecosystem; abstraction for potable supply; abstraction for industrial and agricultural use; water sports; commercial harvesting of fish and shellfish; and special ecosystems. Clearly this scheme will serve a range of different purposes relating to water quality management. However, it is recognised that compliance with water quality parameters of relevant European Community directives is "axiomatic". Notably, the progression towards comprehensive objectives for all waters covered by directives is a long-term enterprise, and the initial exercise has been directed towards surface watercourses rather than groundwater. Meeting specified objectives in practice is likely to be a still more distant prospect, but it is evident that the process which will ultimately lead to specification of statutory objectives for all controlled waters is well under way.

General duties to achieve and maintain objectives

The final part of the national legal mechanism for implementation of Community water directives concerns the relationship between the system of water quality objectives and other powers and obligations of the Secretary of State and the NRA in relation to water pollution. A statutory duty is imposed to ensure that water quality objectives are met and maintained. Hence, it is the duty of the Secretary of State and of the NRA to exercise the powers conferred on them by or under the Act in such a manner as ensures, so far as it is practicable to do so, that the objectives specified for any waters are achieved at all times {s.84(1) WRA 1991}. Most significantly in legal terms, this imposes a *legal duty* upon the NRA to ensure that Community water quality requirements are realised and adhered to at all times.

The practical effect of the legal duty to ensure that water quality requirements are met and maintained is likely to lead to increasingly strict quality requirements being imposed in new discharge consents (see Section 3.4.5). Alternatively, the process of periodic review of discharge consents will be used to realise progressive improvements in discharges to those waters which fail to achieve their specified water quality objective or are in danger of failing to do so. Where a discharge to controlled waters is not the subject of a discharge consent, as may be the case in relation to a highway drain, it may be appropriate for the NRA to utilise other legal powers to restrict the discharge. In some situations it may be necessary to impose a prohibition upon an offending discharge (under {s.86 WRA 1991}, discussed in Section 3.4.1) in order to achieve and maintain a water quality objective.

3.3.3 European Community law and Water Quality Objectives in Scotland

European Community directives concerning water quality impose obligations upon each of the member states. Specifically, in relation to the present study, it has been noted that the Dangerous Substances Directive {76/464/EEC} and the Groundwater Directive {80/68/EEC} will be of particular relevance (see Section 3.3.1). As a member state, the overriding duty upon the United Kingdom Government is to enact national legislation which achieves comprehensive implementation of these directives throughout its territory. In some instances, implementation of Community law may be accomplished by means of a single enactment which applies to England and Wales, Scotland and Northern Ireland. In relation to the water quality directives, however, implementation is achieved by separate, though equivalent, laws within each of the three jurisdictions (enacted under {s.2(2) European Communities Act 1972}).

In respect of the implementation of the water quality directives in Scotland specific obligations will, in most respects, fall upon the national competent authorities, the river purification authorities. In so far as these relate to preventing or authorising the discharge of substances which fall within the lists under the Dangerous Substances and Groundwater Directives, the principal legal mechanisms for achieving this are the system of discharge consents administered by the river purification authorities (see Section 3.4.6) and the power of those authorities to impose prohibitions upon polluting discharges (see Section 3.4.4).

To the extent that Community water directives impose ambient water quality objectives which must be met and maintained in water designated for particular kinds of use, legal provision is made for statutory water quality objectives in Scotland paralleling those measures previously described in relation to England and Wales (see Section 3.3.2). Hence the Secretary of State for Scotland is empowered to prescribe classification systems for various kinds of "controlled waters" (see Section 3.4.2 for the definition of this expression) in Scotland {s.30B COPA 1974}. He may establish water quality objectives for particular waters by reference to a prescribed classification {s.30C COPA 1974}. Thereafter, it will be the legal duty of the Secretary of State and each river purification authority to exercise their powers under Part II of the Control of Pollution Act 1974 and the Rivers (Prevention of Pollution) (Scotland) Acts 1951 and 1965 so as to ensure that, so far as practicable, the water quality objectives specified in relation to particular waters are achieved at all times {s.30D COPA 1974}.

Of specific relevance in relation to the implementation of Community water quality legislation in Scotland are the Surface Waters (Classification) (Scotland) Regulations {SI 1990 No.1121} and the Surface Waters (Dangerous Substances) (Classification) (Scotland) Regulations 1990 {SI 1990 No.126}.

3.4 WATER POLLUTION LAW

3.4.1 Water pollution offences in England and Wales

In relation to the national law relating to pollution from highways, a major consideration must be the extent to which this kind of pollution falls within the traditional criminal offence of water pollution. Most significantly, this offence is committed where a person causes or knowingly permits any poisonous, noxious or polluting matter or any solid waste matter to enter any controlled waters {s.85(1) WRA 1991}.

In certain circumstances, a fishery offence also arises where a person causes or permits any poisonous or injurious liquid or solid matter to be put into waters containing fish or the spawn or food of fish{s.4(1) Salmon and Freshwater Fisheries Act 1975}. However, this overlapping provision is less severely punished than the general offence of water pollution and because of its special character less frequently used. Confining discussion to the general offence of water pollution, a number of observations are of importance in relation to the elements of this offence.

Controlled Waters

First, the expression "controlled waters" determines the kinds of waters to which the offence is to apply. Specifically, controlled waters fall into four subcategories: "relevant territorial waters", "coastal waters", "inland freshwater" and "groundwaters" {s.104 WRA 1991}. In relation to pollution from highways the last two, concerning certain moving and still surface waters and subterranean water, are likely to be of greatest relevance. The meanings of "inland freshwater" and "groundwaters" are determined by the following chain of definitions.

- "Inland freshwater" means the waters of any relevant lake or pond or of so much of any relevant river or watercourse as is above the fresh-water limit {s.104(1)(c) WRA 1991}. "Relevant lake or pond" means any lake or pond which, whether it is natural or artificial or above or below ground, discharges into a relevant river or watercourse or into another lake or pond which is itself a relevant lake or pond {s.104(3) WRA 1991}. "Watercourse" includes all rivers, streams, ditches, drains, cuts, culverts, dykes, sluices, sewers and passages through which water flows except mains and other pipes which belong to the NRA or a water undertaker or are used by a water undertaker or any other person for the purpose only of providing a supply of water to any premises {s.221(1) WRA 1991}. "Relevant river or watercourse" means any river or watercourse, including an underground river and an artificial river or watercourse, which is neither a public sewer nor a sewer or drain which drains into a public sewer {s.104(3) WRA 1991}. "Drain" is stated to have the same meaning as in the Water Industry Act 1991. That Act defines the expression to mean a drain used for the drainage of one building or of any buildings or yards appurtenant to buildings within the same curtilage {s.219(1) WIA 1991}.

- "Groundwaters" are defined as any waters which are contained in underground strata {s.104(1)(d) WRA 1991}. "Underground strata" means strata subjacent to the surface of any land {s.221(1) WRA 1991}.

Causing or knowingly permitting

Given that a discharge of poisonous noxious or polluting matter from a highway is made to an inland freshwater or groundwater, the second ingredient is that a guilty person must be shown to have *caused or knowingly permitted* the entry of the matter into the controlled waters. The precise meanings of "cause" and "knowingly permit" have occupied the attention of the courts on many occasions. From the outset it is established that the expressions have distinct meanings and that it is sufficient for the prosecution to establish that *either* the accused caused pollution *or* that he knowingly permitted it for a conviction to succeed. That is, *both* elements need not be shown.

Summarising an extensive body of case-law concerned with the extent of the offence as it was provided for under previous legislation, it has been established that causing water pollution is an offence of "strict liability", sometimes loosely referred to as an "absolute offence". The leading decision of the House of Lords in *Alphacell Ltd. v Woodward* in 1972 {[1972] 2 *All England Law Reports* 475} establishes that "cause" does not require an intention to pollute waters or negligence on the part of the polluter to be shown.

A pertinent example of the operation of strict liability in this context is provided by the decision in *Wrothwell Ltd. v Yorkshire Water Authority* (*Criminal Law Review*, 1984) where a director of a company deliberately poured 12 gallons of a concentrated herbicide down a drain. It was known that the herbicide was toxic to fish life, but the natural expectation was that the liquid would pass down the drain into the public sewer system. In fact, the company's drain did not connect with the public sewer but to a system of pipes leading to a nearby stream where the eventual discharge caused a substantial fish-kill. The decision of the court was that, even though the director may not have intended the water pollution incident which took place, it was nonetheless the consequence of his action and, therefore, he was guilty of "causing" the pollution within the meaning of the offence.

Guilt in relation to the water pollution offence, may be established *either* by showing that the accused caused the entry or discharge *or* that he knowingly permitted it. "Knowingly permit" in this context means that the accused failed to prevent the entry or discharge of the matter concerned into controlled waters when it was within his power to do so, accompanied by knowledge that the discharge or entry was taking place.

A recent illustration of the interpretation "knowingly permit" in this context is provided by the decision in *Schulmans Incorporated Ltd. v. National Rivers Authority* {unreported, 3 December 1991, Queen's Bench Division}. The facts giving rise to the prosecution concerned a spillage of fuel oil near a tank on the accused's premises which found its way into the drainage system on its land, and from there into a nearby watercourse. On appeal, the court found that the company should be acquitted because it had not been satisfactorily established that it had knowingly permitted the escape of fuel oil and consequent water pollution. In particular, there was no evidence that the company could have taken preventative action more swiftly than it did, or that there was an escape of oil which it could have prevented but failed to prevent. Accordingly, they had not been shown to have *permitted* the entry of the oil into the watercourse within the wording of the charge.

Contravention of Prohibitions

As will be seen, the general water pollution offence is subject to a particular defence relating to highway drains {under s.89(5) WRA 1991}. The effect of this exception is that, ordinarily, no offence will be committed by a highway authority in causing or permitting a discharge to be made from a drain kept open in accordance with highways legislation {s.100 HA 1980}. However, this general defence can be lost in circumstances where the NRA specifically imposes a prohibition upon the discharge. Hence a principal water pollution offence is committed where a person "causes or knowingly permits any matter, other than trade effluent or sewage effluent, to enter controlled waters by being discharged from a drain or sewer in contravention of a prohibition" {s.85(2) WRA 1991}.

It is to be noted that the offence arising through contravention of a prohibition applies in relation to "any matter, *other than* trade effluent or sewage effluent" since these types of effluent are separately provided for {s.85(3) WRA 1991}. As the definitions of trade and sewage effluent specifically exclude surface water, including water from roofs {s.221(1) WRA 1991}, the offence can be committed, where a prohibition is imposed, despite the fact that the matter concerned is surface water such as runoff from highways.

However, the overriding consideration is that the offence arises in respect of matter entering controlled waters by being discharged from a drain or sewer "in contravention of a prohibition". This relates to the power of the NRA specifically to prohibit certain discharges by notice or regulations {under s.86 WRA 1991}. This may be done where the NRA gives a person notice prohibiting him from making or continuing a discharge, or from making or continuing a discharge unless specified conditions are observed {s.86(1) WRA 1991}. Alternatively, a discharge may be in contravention of a prohibition if the matter discharged contains a prescribed substance, or derives from a prescribed process or from a process involving the use of prescribed substances or the use of such substances in quantities which exceed prescribed amounts {s.86(2) WRA 1991}.

It may be observed that these powers are broadly formulated so that the NRA could apply prohibitions in relation to highway drains or sewers if it saw fit to do so because of some particular pollution hazard, and, thereafter, the continued use of the drain or sewer would amount to the offence of breach of the prohibition {under s.85 WRA 1991} and the general defence for highway drains would not be available.

Another important consequence of the application of a prohibition in relation to a highway drain, and subsequent breach of that prohibition, is that this will empower the NRA, unilaterally, to serve a notice of consent in respect of the discharge {under Schedule 10 para.5(1) WRA 1991}. Discharge consents, the consequences of their breach and rights of appeal against the determination or imposition of a consent, are discussed later (see Section

3.4.5). Curiously, there appears to be no corresponding statutory right of appeal against a decision of the NRA to impose a prohibition upon a highway drain, despite the fact that a prohibition imposed subject to conditions may serve a similar practical purpose to the imposition of a discharge consent.

3.4.2 Water pollution offences in Scotland

Whilst Part II of the Control of Pollution Act 1974 {ss.31 to 56} remains the principle enactment concerning water pollution in Scotland, this Part of the Act was substantially revised by provisions substituted under Schedule 23 to the Water Act 1989, and further amendments have been brought about under the Environmental Protection Act 1990 and the Natural Heritage (Scotland) Act 1991.

As amended, Part II of the Control of Pollution Act 1974 applies to "controlled waters" which are defined as encompassing "relevant territorial waters", "coastal waters", "inland waters" and "groundwaters" {ss.30A and 56 COPA 1974}. With some minor variations in wording, these terms are defined so as to correspond with the same types of waters as are "controlled waters" in England and Wales (under {s.104 WRA 1991}, see Section 3.4.1). The Secretary of State may by order provide that certain waters which do not come within the definition of controlled waters are nevertheless to be treated as controlled waters ({s.30A(5) COPA 1974}, and see Controlled Waters (Lochs and Ponds) (Scotland) Order 1990, {S.I.1990 No.120}).

The key water pollution offences arise if a person causes or knowingly permits any poisonous, noxious or polluting matter to enter controlled waters {s.31(1) COPA 1974}, or if a person causes or knowingly permits any trade or sewage effluent to be discharged into any controlled waters {s.32(1) COPA 1974}. Distinct charges will need to be brought in respect of "knowingly permitting" and "causing" water pollution, and the latter has been held to be an offence of strict liability ({*Lockhart v. National Coal Board* (1981) *Scottish Criminal Cases Reports* 9}, and see Section 3.4.1).

3.4.3 Defences in England and Wales

Defences and authorised discharges

In England and Wales the offence of causing or knowingly permitting the pollution of controlled waters is subject to a number of stated defences and exceptions provided for under the Water Resources Act 1991, and is also subject to a system of authorisations for discharges to controlled waters including "discharge consents". Confining the discussion to those defences which may be especially relevant to pollution from highways, two matters are particularly pertinent: a defence provided for in respect of emergencies, and a special provision relating to highway drains.

Emergencies

With regard to emergencies, a person will not be guilty of the water pollution offence in respect of an entry of polluting matter into controlled waters if three factors are established. First, the entry is caused or permitted, or the discharge is made, in an emergency in order to avoid danger to life or health; second, that person takes reasonably practicable steps for minimising the extent of the entry or discharge and of its polluting effects; and, third, that the particulars of the entry or discharge are furnished to the NRA as soon as reasonably practicable after it occurs {s.89(1) WRA 1991}.

Notably, this defence might become available in circumstances where a spillage of a hazardous substance occurred onto a highway as a result of a road accident involving a vehicle carrying the substance. The effect of the defence would be that if emergency services personnel sought to wash the polluting substance away in order to avoid danger to life or health, and this resulted in the substance entering controlled waters causing pollution of surface water or groundwater, the emergency services personnel would have a defence to the water pollution offence provided that they caused the minimum pollution possible and that particulars of the incident were furnished to the NRA as soon as reasonably practicable afterwards.

Two observations are warranted in relation to the emergency defence. First, the defence is stated to be available where action is taken "in order to avoid danger to life or health". This expression is not defined, and it is a matter of speculation as to whether it should be confined to *human* life or health or to extend to concerns for non-human life or health, as where farm livestock are threatened, or indeed whether harm to living creatures in the natural environment should also be taken into account in taking emergency action.

Second, although the emergency defence would generally be available to the emergency services when taking action which involves causing or permitting the entry of polluting matter into controlled waters in order to alleviate an emergency, the availability of the defence to the person who has brought about an emergency is less likely. Hence, if a tanker driver is responsible for causing the circumstances leading to the entry of the polluting matter into a watercourse, it is suggested that the emergency defence would not be available to him (see *Waste Incineration Services Ltd. v. Dudley Metropolitan Borough Council* [1992] *Environmental Law Reports* 29, and *Perka v. The Queen* [1985] 13 *Dominion Law Reports* (4th) 1). Clearly, this raises immensely difficult issues of causality as, for example, where it is not apparent who is to blame for a road traffic accident which results in water pollution. Nevertheless, as a general principle, the emergency defence must be limited in its availability to those persons who take ameliorative action in an emergency and will not be available to those whose acts precipitate the emergency.

Highway Drains

A highway authority may, for the purpose of draining a highway, construct or lay drains in the highway or in land adjoining it. Similarly, it may erect barriers in the highway and adjoining land to divert surface water into or through drains, and scour, cleanse and keep open all drains in the highway or adjoining land ({s.100(1) HA 1980}, see Section 3.2.35). Where these powers are exercised, a special defence is given to the highway authority in criminal proceedings where water pollution arises from discharges made through highway drains which are kept open by virtue of these powers.

Specifically, where a highway authority or other person is entitled to keep open a drain, by virtue of s.100 of the Highways Act 1980, that person will not be guilty of the offence of water pollution by reason of his causing or permitting any discharge to be made from the drain, unless the discharge is made in contravention of a prohibition imposed under s.86 of the Water Resources Act 1991 {s.89(5) WRA 1991}. That is to say, a general defence exists to the water pollution offence relating to discharges from highway drains, but this defence may be overridden through prohibition by notice or regulations as previously described (see Section 3.4.1) or the imposition of a discharge consent as described below (see Section 3.4.5).

It is to be noted that the defence to the water pollution offence is only available to the highway authority or the person lawfully entitled to keep open the drain. The counterpart of this is that the defence could not be claimed by another person making illegitimate use of a highway drain for the disposal of a substance which subsequently resulted in pollution of controlled waters (see *National Rivers Authority v. Tarmac Construction Ltd.* [1993] *Environmental Law Brief* 19). Hence, for example, if a person with no rights in respect of a highway drain was to use it for the disposal of waste oil, resulting in surface or groundwater contamination, the defence relating to highway drains would not be available.

Where a highway authority constructs a highway drain, as described above, the water may be discharged directly, or through an existing drain, into any natural or artificial inland waters or any tidal waters {s.100(2) HA 1980, and s.299(1) HA 1980}. Where a highway authority exercises this power so as to interfere with any watercourse or drainage works vested in or under the control of the NRA, or any other drainage body or navigation authority, the consent of the body or authority concerned is required. However, this consent may not be unreasonably withheld, and any dispute as to the reasonableness of it being withheld may be referred to arbitration {s.339 HA 1980}.

The highway authority, in exercising any of the powers to construct or maintain drains, will be bound to pay compensation to the owner or occupier of any land who suffers damage through the construction or maintenance of the drain or the discharge of water through it {s.100(3) HA 1980}. The precise meaning of "damage" in this context is not statutorily defined, however, and gives rise to some uncertainty as to the kinds of impairment which might require compensation. On a narrow interpretation "damage" might be taken to encompass only those losses incurred by the activities involved in the actual construction of the discharge point in the receiving waters and losses directly incurred by drain cleaning operations. On a broader view it might be taken to include continuing forms of damage to the receiving waters resulting from either the quantity or quality of water discharged through the drain. Disputes as to the amount of compensation payable are to be assessed, if the parties agree, by arbitration or, in cases of disagreement, by the County Court {s.308 HA 1980}.

Notably also, the existence of a defence to the criminal offence of water pollution will not provide any defence to civil proceedings arising in respect of pollution of a watercourse originating from highway runoff ({s.100 WRA 1991}, see Section 3.5.3). Accordingly, in one instance a highway authority was found to be civilly liable in respect of water, fouled with tar from the surface of a road, which entered a watercourse {*Dell v. Chesham Urban District Council* [1921] 3 *King's Bench Reports* 427}. By contrast, in another case, a highway authority was found not to be civilly liable where water carrying sand and silt entered a watercourse where this had no significant effect upon the quality of the receiving waters {*Durrant v. Branksome Urban District Council* [1897] 2 *Chancery Law Reports* 291}.

3.4.4 Defences and authorisations in Scotland

In Scotland a defence is provided for in relation to the water pollution offence if the entry of polluting matter is authorised by, or is a consequence of an act authorised by, a consent given by the Secretary of State or a river purification authority under COPA 1974 and the entry or act is in accordance with the conditions, if any, to which the consent is subject ({s.31(2)(a) COPA 1974}, and see Section 3.4.6 on discharge consents).

An emergency defence is provided for where a person is not guilty of the water pollution offence where the entry is caused or permitted in an emergency in order to avoid danger to life or health. This defence is only available, however, where the person can claim that all reasonably practicable steps were taken to minimise the extent of the entry and its polluting effects and, as soon as practicable after the entry occurs, particulars of it are furnished to the appropriate river purification authority {ss.31(2)(c) and 32(4)(b) COPA 1974}.

Another defence is provided for in respect of matter other than trade or sewage effluent discharged into controlled waters from a drain which a roads authority is obliged or entitled to keep open {by virtue of s.31 RSA 1984}. However, this defence will not be available where the appropriate river purification authority has served a notice of prohibition upon the roads authority relating to the drain. For these purposes this is to give at least three months notice of the prohibition to the roads authority. Where a notice of prohibition is imposed in relation to a road drain, it will be an offence to discharge any matter into controlled waters from the drain unless the discharge is made under a discharge consent {under s.34 COPA 1974} and it is in accordance with the conditions, if any, to which the consent is subject {ss.31(2)(d) and 32(1)(c) COPA 1974}. It appears that no right of appeal is provided for in relation to the imposition of a prohibition notice, though rights of appeal exist in respect of a discharge consent imposed consequent upon the imposition of a notice or prohibition, described below.

Curiously, the offence in relation to discharges into and from road drains may be committed by *any person* where a notice of prohibition is imposed. Conversely, the defence in relation to road drains will be available to any person, and not merely the roads authority concerned {s32(1)(c) COPA 1974}.

3.4.5 Discharge consents in England and Wales

In relation to highway authorities or other persons entitled to keep open a highway drain in England and Wales under the Highways Act 1980, the most appropriate defence to the water pollution offence is the explicit provision relating to highway drains ({under s.89(5) WRA 1991}, see Section 3.4.3). Where contaminated water from a highway is discharged otherwise than through a highway drain, or otherwise than by a highway authority, this defence will not be available. Also, as has been explained, the imposition of a prohibition upon a highway authority {under s.86 WRA 1991} and, in some circumstances, the imposition of a discharge consent ({under Schedule 10 para.5 WRA 1991}, and see below) may result in the loss of the defence relating to highway drains.

Generally, important practical exceptions to the water pollution offence arise where a person has an authorisation of a specified kind which legitimises a discharge which would otherwise amount to a water pollution offence. Although various kinds of authorisation may serve this purpose, the most important are "discharge consents". Thus, it is provided that a person will not be guilty of the water pollution offence, in respect of the entry of matter into any controlled waters or any discharge, if the entry occurs or the discharge is made under and in accordance with a discharge consent provided under the Water Resources Act 1991 {s.88(1)(a) WRA 1991}.

Procedural requirements in relation to discharge consents normally require applications for consents to be accompanied or supplemented by all such information as the NRA may reasonably require. Notice of an application is to be published by the NRA and copies of the application sent to every local authority or water undertaker within whose area the proposed discharge is to occur, unless the discharge will have no appreciable effect on the waters into which the discharge is to be made. The NRA is then to consider any written representations or objections to the application made within a specified period. Consents may be granted subject to any conditions the NRA may think fit and may include conditions such as to the nature, origin, composition, temperature, volume and rate of the discharges and as to the periods during which the discharges may be made {Schedule 10 WRA 1991}.

The NRA is under a duty to review consents from time to time, and the outcome of this process may be the revocation of a consent, or the modification of its conditions, or the making of a previously unconditional consent subject to conditions. Modification of consents can also be made by the NRA as a result of a direction given by the Secretary of State. Where a consent is given by the NRA for a discharge, the instrument signifying the consent is to specify a period during which no notice of revocation or modification will be served in relation to the consent. The period during which the revocation or variation of a consent is precluded, without the consent of the person making the discharge, is to be a period of not less than two years. Although it is legally possible that variation of a consent may be undertaken within the two year minimum period for review, it is likely that this procedure will give rise to a duty upon the NRA to compensate the discharger for any loss or damage sustained as a consequence of the variation of the consent {Schedule 10 WRA 1991}.

Where, on application or review, a discharge consent is refused, revoked, modified or made subject to conditions which are unacceptable to the applicant or holder of the consent, a right of appeal is provided for {s.91 WRA 1991}. Similarly, where the NRA imposes a consent upon a highway authority, in the circumstances described below, the same right of appeal exists. This allows the applicant, or the person who seeks to make the discharge, to appeal to the Secretary of State against the adverse decision of the NRA {s.91(2) and (3) WRA 1991}. On appeal, the Secretary of State is empowered to direct the NRA to modify or reverse its decision. Accordingly, he may require the consent to be given unconditionally or subject to specified conditions {s.91(5) WRA 1991}.

Discharge consents for highway drains

Under most circumstances discharge consents are not required in respect of highway drains, since a consequence of the explicit exception to the water pollution offence for highway drains {under s.89(5) WRA 1991} is that no offence is committed whether or not a particular drain is the subject of a consent. However, there is an important exception to this which may arise if a highway drain becomes the subject of a notice of prohibition {under s.86 WRA 1991}. Where a highway drain is the subject of a prohibition notice and effluent has been discharged in contravention of the prohibition, and a similar contravention is likely, the NRA may serve a discharge consent upon the person who has caused or permitted the contravention of the prohibition {Schedule 10 para.5(1) WRA 1991}. Thereafter, breach of the consent will amount to a criminal offence {see s.85(6) WRA 1991}.

The imposition of discharge consents "without application" may not relate to discharges which have taken place before notice of the consent was served. Otherwise, the general requirements in relation to discharge consents apply similarly as for consents which are granted following the application procedure described above. Hence, the NRA may impose the same conditions in the consent, and it will be subject to corresponding duties in relation to publicity and consideration of representations, as apply in respect of prospective applications for consents. Likewise, a discharge consent which is imposed without application may be the subject of an appeal to the Secretary of State as described above.

3.4.6 Discharge consents in Scotland

In Scotland applications for discharge consents are normally made to the appropriate river purification authority and are to be accompanied by all information reasonably required by the authority {s.34(1) COPA 1974}. The river purification authority is to publish details of the application, send details to each local authority in whose area the proposed discharge is to be made and is to consider representations relating to the application {s.36 COPA 1974}. The Secretary of State may direct a river purification authority to transmit an application to him for determination, and he may cause a local inquiry to be held into the application {s.35 COPA 1974}. Appeals may be made to the Secretary of State where it is alleged that a river purification authority has unreasonably withheld its consent or imposed unreasonable conditions in a consent {s.39 COPA 1974}. Duties are imposed upon river purification authorities to review consents from time to time, though certain restrictions are imposed upon the time period within which a consent may be varied or revoked by an authority {ss.37 and 38 COPA 1974}.

Of special significance in relation to road drains is the power of river purification authorities to impose discharge consents on drains which are the subject of a notice or prohibition. Hence, if a person has caused or permitted matter to be discharged contrary to a prohibition upon a road drain, and a similar contravention by that person is likely, the river purification authority may serve an instrument on him giving its consent, subject to conditions, for discharges of a specified kind {s.34(3) COPA 1974}. Thereafter, no offence will be committed providing an entry of matter is authorised by, and in accordance with the conditions of, the consent {ss.31(2)(a) and 32(1)(c) COPA 1974}. A person upon whom a consent is imposed, or upon whom a river purification authority unreasonably declines to impose a consent, or upon whom a consent containing terms which are alleged to be unreasonable is granted may appeal to the Secretary of State {s.39 COPA 1974}.

3.5 WATER RESOURCE PROTECTION

3.5.1 Water resources and the National Rivers Authority

Whilst the pollution of surface water or groundwater by runoff from highways may give rise to criminal proceedings in the circumstances that have been described, another possibility to be considered is that highway drainage may result in the contamination of surface or subterranean waters that are utilised as a source of supply for domestic or other purposes. Naturally, this possibility will be of concern to water services companies and statutory water supply companies in England and Wales. Equally, the NRA is bound to have regard to the water supply duties imposed upon suppliers in exercising its powers ({s.15(1) WRA 1991}, and see below).

Water supply duties arise under the Water Industry Act 1991, and require every water undertaker to develop and maintain an efficient and economical system of water supply within its area and to ensure that arrangements have been made for providing supplies of water to premises in its area {s.37(1) WIA 1991}. In respect of water quality, a duty exists to supply water to domestic premises which is "wholesome", as this expression is defined under regulations made by the Secretary of State {ss.67 and 68 WIA 1991, and Water Supply (Water Quality) Regulations SI 1989 No.1147 as amended}. It is an offence for a water undertaker to supply water, by means of pipes to any premises, which is unfit for human consumption {s.70(1) WIA 1991}. Accordingly, the NRA will be bound to have regard to this particular obligation upon water suppliers when exercising its powers.

The protection of water resources by conserving, redistributing or otherwise augmenting such resources is amongst the principal functions of the NRA {s.19 WRA 1991}. It follows that where water resources are provided by either surface waters or aquifers, there is a shared concern on the part of the NRA and water undertakers that these supplies are free from contamination due to runoff from highways.

This co-incidence of interests is reinforced by an obligation upon the NRA, in exercising any of its powers, to have particular regard to the water supply duties, imposed under the Water Industry Act 1991 and described above, on any water undertaker which appears to be or to be likely to be affected by the exercise of the power in question {s.15(1) WRA 1991}. Thus, for example, in granting a discharge consent to surface or groundwater, the NRA will be bound not only to have regard to the protection of the aquatic environment in general, but also to the ultimate effect of the proposed discharge upon the "wholesomeness" of water which may be used for water supply purposes.

The main basis through which the NRA fulfils its function of conserving, redistributing or otherwise augmenting water resources, is by an abstraction licensing system which encompasses situations where water abstraction is sought by a water undertaker {s.167 WIA 1991}. The general effect of a licence to abstract water is that the holder is to be taken to have the right to abstract water to the extent authorised by the licence and in accordance with the provisions contained in it. Accordingly, where any legal action is brought against a person in respect of water abstracted from a source of supply, it will be a defence for that person to prove that the abstraction was in pursuance of a licence and the provisions of the licence were complied with {s.48(2) WRA 1991}.

A pertinent issue concerns the relationship between a licensed abstractor of water from a surface or underground source and the owner of a highway, the discharge from which is alleged to have an adverse effect upon the quality or quantity of water which is authorised to be abstracted. Generally, it is provided that the possession of an abstraction licence does not confer any special right of action in civil proceedings nor derogate from any right of action in civil or criminal proceedings {s.70 WRA 1991}. This means that the possession of an abstraction licence would not provide the licence holder with any special civil rights as against the owner of the highway which was alleged to have caused contamination of the source of supply from which abstraction was made. Likewise the licensee would not be impeded in seeking civil redress by the possession of the abstraction licence. Hence, the possession of an abstraction licence will neither aid nor hinder a civil action involving the contamination of a water supply {*Cargill v. Gotts* [1981] 1 *All England Law Reports* 682}.

Similar observations can be made in relation to the defence to criminal proceedings which is allowed to a highway authority in respect of discharges from highway drains ({s.89(5) WRA 1991, s.100 HA 1980} and see Section 3.4.3). Whilst no criminal offence is ordinarily committed in respect of discharges from highway drains, the same defence does not apply in relation to civil proceedings for water pollution originating from highway drains. Indeed, it is expressly stated that none of the water pollution provisions under the Water Resources Act 1991 derogates from any right of action or other remedy, whether civil or criminal, in proceedings instituted otherwise than under the Act ({s.100 WRA 1991}, and see Section 3.5.3 on the civil law).

3.5.2 Water resource protection in Scotland

Public water and sewerage services in Scotland are presently provided by local authorities consisting of the nine regional councils and three island councils. However, the means of financing and managing these services are presently under review and options for the restructuring of the Scottish water industry, including various mechanisms for private investment, are set out in the Scottish Office consultation paper *Water and Sewerage in Scotland Investing for Our Future* (1992).

The regional and island councils act as water supply authorities under the Water (Scotland) Act 1980, and also as the providers of sewerage services under the Sewerage (Scotland) Act 1968. As water authorities, the councils have a duty to provide a "wholesome" supply of piped water where it is required for domestic use unless this cannot be done at reasonable cost (see the Water Supply (Water Quality) (Scotland) Regulations 1990, SI 1990 No.119). Duties in respect of sewerage require water authorities to provide public sewers to drain domestic sewage, surface water and trade effluent and to treat or otherwise deal with the contents of sewers.

Notably, the river purification authorities are not placed under any duty, in exercising their powers, to have regard to the duties imposed upon water supply authorities or the providers of sewerage services. This contrasts with the duty imposed upon the NRA in England and Wales to have regard to the duties imposed upon the water industry (contrast {s.115 WRA 1991} discussed in Section 3.5.1).

As has been noted, the position of river purification authorities in Scotland differs from that of the NRA in that the former lack the range of functions relating to the protection of the aquatic environment possessed by their counterpart in England and Wales. Whilst the river purification authorities are under a general duty to promote the cleanliness of waters in Scotland {s.1(1) R(PP)(S)A 1951}, and to use their particular powers for that purpose, they are not placed under any specific obligation to have regard to the water supply duties imposed upon local authorities {under the W(S)A 1980}.

Although a general duty is placed upon the Secretary of State for Scotland to promote the conservation of water resources in Scotland {s.1(a) Water (Scotland) Act 1980}, this duty will relate only indirectly to certain situations where particular water resources are threatened by contaminating activities. More direct provision is provided for by way of byelaw creating powers given to water authorities, which allow for the imposition of special controls to prevent pollution of water within the catchment area and to prohibit or regulate activities which have that consequence {s.71 W(S)A 1980}.

3.5.3 The Civil Law

Whilst the preceding discussion has focused upon matters of public regulatory law relating to pollution from highways, primarily concerning criminal offences and licensing provisions, it is essential to discuss a distinct legal aspect of water pollution: the civil law. Broadly, the civil law may be characterised as the collection of private legal rights and duties which arise between individuals and corporate bodies and by which a person may be held liable for a recognised kind of harm inflicted upon another. Although the criminal provisions which have been discussed rest upon punitive sanctions, the civil law is primarily concerned to provide redress for harms of various kinds. The precise kind of redress which is available will depend upon the circumstances at issue, but generally an individual who seeks to bring a civil action will be seeking a compensatory award of damages for the loss which has been suffered, or a court order to prevent the continuation of the offending activity which is at issue.
Exaggerating somewhat, the contrast is that the criminal law is essentially *punitive* in character whilst the civil law is essentially *compensatory*.

In relation to water, it has to be recognised that important private rights exist in watercourses which, if infringed, enable the owner of the watercourse to bring a civil action against the offender. The rights of owners of watercourses are termed "riparian rights" and the law protects these rights in so far as they relate to interests in water quality, water quantity and certain ancillary matters such as fisheries and navigation. Accordingly, where a discharge of contaminated water into a watercourse is the subject of complaint, a downstream riparian owner will be entitled to a remedy where he can show that this has caused an interference with his riparian interest in the watercourse. The fundamental principle was stated in a leading decision, "Every riparian proprietor is thus entitled to the water of his stream in its natural flow, without sensible diminution or increase, and without sensible alteration in its character or quality. Any invasion of this right causing actual damage . . . entitles the party injured to the intervention of the court" {*John Young and Co. v. Bankier Distillery Co.* (1893) [1891-4] *All England Law Reports* 439, per Lord Macnaghten at p.441}. It follows that where "sensible alteration" to the quality of a watercourse can be shown, as a result of contamination originating from highway runoff, an action may, in principle, be brought for compensation for the loss suffered (see Section 3.4.3). Although riparian rights, strictly so called, do not exist in water which is present in underground strata, it is also possible that a right to redress may arise for contamination of this kind of water, under certain circumstances.

The branch of the civil law which is most relevant to water pollution actions is termed the "tort of nuisance" in England and Wales, and the "delict of nuisance" in Scotland. Further terminological differences exist between the two jurisdictions in that a court order to prevent the continuation of a particular nuisance is called an "injunction" in England and Wales and an "interdict" in Scotland. Despite these variations in nomenclature, however, the general principles of civil liability for water pollution are essentially the same in both jurisdictions {*Flemming v. Hislop* (1886) 13 *Session Cases 4th Series (Rettie)* (House of Lords) 43, per Lord Fitzgerald at p.48}, though some contrasts may arise due to the different interests in land which may exist in Scottish law.

Thus, the basic principle of civil law in England and Wales, that a riparian owner is entitled to a remedy where sensible alteration is caused to the quality of his watercourse, has a direct counterpart in Scottish law. This is that, in relation to a private stream, that is a stream which is non-navigable, riparian owners are entitled to use the water subject to the condition that it is passed down to lower neighbours undiminished in quantity and unimpaired in quality by the addition of any unnecessary or artificial impurity {*Duke of Buccleuch v. Alexander Cowan and Sons* (1866) 5 *Session Cases 3rd Series (Macpherson)* 214}. Similarly, whilst riparian rights, strictly so called, arise only in relation to surface and underground water flowing in a defined channel, certain interests in respect of percolating water passing through the ground by an undefined route may be protected in civil law in Scotland. Hence, if the owner of higher land causes the contamination of water which passes by percolation to lower land, the degradation of water quality may form the basis of an action by the lower landowner {*Irving v. Leadhills Mining Co.* (1856) 18 *Session Cases 2nd Series (Dunlop)* 833}.

The Cambridge Water Company Case

Recently, the potential for civil litigation in relation to the pollution of water supply sources has been authoritatively clarified by the decision of the House of Lords in the case of *Cambridge Water Company v. Eastern Counties Leather PLC* {House of Lords, 9 December 1993}. Briefly, the facts were that tetrachloroethane was discovered in tap water supplied by the Cambridge Water Company in amounts exceeding those permitted under the EC Directive on Drinking Water {89/778/EEC}. The solvent was traced to a particular borehole and found to have originated from the defendant's leather tannery, though contamination had resulted from spillages which had occurred some years previously. The water company sued the defendants for compensation amounting to approximately £1m, a sum which it had been obliged to spend in opening up a new borehole to secure a replacement supply.

The House of Lords found that the water company was not entitled to recover the damages sought. The basis for this decision was that the foreseeability of harm was an essential element in establishing civil liability in these circumstances. Since at the time the spillages of solvent took place, before 1976, it could not be foreseen that they would lead to contamination of water supplies under an EC directive that had not yet been enacted, no liability could arise against the leather tannery.

Some care must be taken in the interpretation of the *Cambridge Water Case*. Notably the Court was concerned not to impose an unreasonably high standard in relation to "historic" pollution. Implicitly it would be wrong to impose liability upon present day landowners for past activities, the polluting potential of which was not appreciated at the time they occurred. On the other hand it would be no defence in a civil action for a modern polluter to claim that he did not know, or could not foresee, the contaminating potential of chemical spillage upon water to be used for supply purposes. The key point is that the foreseeability of harm must be judged at the time when the offending activity took place and not in the light of later knowledge of the adverse environmental effects of the activity.

Relating this back to the problem of pollution from highways, if it is established that surface water runoff from highways is known to have a potentially contaminating effect upon surface or groundwater sources of supply, then the potential for strict civil liability arises. As has previously been noted, although a highway authority, from whose road the contamination has originated, may have a defence to any criminal proceedings which may be brought, this will not avail the authority in respect of civil proceedings {s.100 WRA 1991} and, as the recent decision illustrates, the magnitude of such claims may be considerable.

4 Current practice and control options

4.1 THE FUNCTION OF HIGHWAY DRAINAGE

The principal function of a highway drainage system is to remove surface water quickly from the road surface thereby reducing the risk of accidents. A secondary function is to provide adequate drainage of the road foundations and construction layers. This reduces the maintenance required and prolongs the life of the road.

Pollutants can enter a highway drainage system from a variety of sources. These sources, and the impact they may have on receiving waters, are described in detail in Section 2.

Control of pollution discharged from highway drainage can be achieved through: appropriate design and careful construction of new systems; rehabilitation or modification of existing systems; and by pollution management, for example through control of maintenance operations.

This Section describes the general principles of highway drainage design in relation to the various types of systems in common use and the effects of these systems on water quality. Other methods of controlling pollution from highway drainage systems are also described.

4.2 GENERAL DESIGN CONSIDERATIONS

4.2.1 Runoff

Highway surfaces have very low permeability in general, although verges and central reservations etc. are likely to be more permeable. Some papers (e.g. Bellinger et al, 1982) refer to an impermeability factor, a range for which is given in British Standard BS 8301 as 0.75-1.0 (meaning that 75-100% of rainfall runs off). The general use of such figures can be misleading except perhaps in intense storm conditions, when high runoff percentages do tend to occur. It has been estimated (Ellis et al, 1986) that 0.1-1.5mm of rain is necessary to initiate runoff. Runoff can be restricted in urban areas by blocked drains (e.g. by leaf falls in autumn). Ellis et al (1986) have estimated runoff figures to be between 34% and 83% for an urban residential setting in London, which probably includes all of the above factors.

Rainfall/runoff relationships therefore depend, inter alia, on:

- depth, duration, frequency and pattern of rainfall
- surface conditions - roughness, permeability, slopes
- exposure of site
- condition of drainage system.

This means that relationships will tend to be site and event specific, making predictions of runoff quantity difficult.

4.2.2 Design criteria and techniques

As this report is concerned with discharges from roads, the design criteria adopted by the Department of Transport and its Agents are considered in detail.

Highway drainage design is based upon the recommendations of Road Note 35 (DoE/TRRL, 1976). This allows for designs to be prepared using the Rational (or Lloyd Davis) Method (HR Wallingford, 1986) and also for computer analysis using hydrograph manipulating programs such as HECB Hydran (DOT, 1991; HR Wallingford, 1986). The recommended procedure is for the initial design to be prepared on peak flow calculations from Lloyd Davis (using hand calculation or one of the many computerised versions produced for PCs) followed by a detailed analysis using a hydrograph manipulative program to assess the performance of the preliminary design. In this latter process, typically, an input hyetograph is routed through a paved catchment using a time/area method whilst at the same time a simple linear reservoir method is used for unpaved contributory areas. The choice of hyetograph is critical to such analysis and therefore sensitivity testing should always be carried out for the two fundamental properties of the system, i.e. peak flow capacity and attenuation.

Whichever method is used there are five basic criteria that must be met by the final design:

- pipes are designed to run full at peak flow for a design storm with a probability of a once in one year occurrence
- the system must be adequate to drain water from a design storm with a probability of a once in five years occurrence without water standing on the carriageway. In practice this means that a check has to be carried out on the system for the surcharged condition
- where attenuation storage is to be provided to restrict the rate of discharge into a stream, then rainfall events in excess of a 1:10 year period are unlikely to be contained by the drainage/road structure. Excess water will be lost to adjacent land
- minimum velocity in pipes should ideally be >0.8 m/s in part full pipes for a 1:6 month event. Maximum velocity should not exceed 3.1 m/s unless special provision is made
- minimum pipe size for conveyance systems should be 150mm diameter with a normal upper limit of 900mm diameter (DMRB Vol 4, HA40/89). Pipes in excess of this diameter may be used but will be subject to structural assessment and approval procedures.

Other variables such as pipe roughness co-efficient, choice of design storm for attenuation storage calculations, time of entry and impermeability co-efficient will be decided by the designer and client so as to achieve the required level of service at the most economic cost.

For guidance purposes the most commonly used forms of storm hyetograph for hydrograph analysis are: a 75 percentile winter storm for storage calculations and a 50 percentile summer storm for peak flow capacity analysis.

4.2.3 Pollution

Section 2 deals in detail with the types of pollution arising from highway drainage. In this and the following sections the range of methods for dealing with highway runoff is discussed and particular attention is focused on the capability of the methods to contribute to the reduction of the pollutant impact on the receiving body of water. It is important to emphasise that reduction of the impact of pollution generated by traffic and road use need not depend solely on treatment at the point of discharge. Individual components of a drainage system can make significant contributions to improvements in the quality of the discharge, at little or no additional cost, although they may not initially have been selected for that function.

The present perception of highway runoff, amongst drainage engineers, is that quantity constitutes a greater problem than quality. Conversely to the water regulatory authorities, highway runoff constitutes a threat to the quality of the water environment, particularly underground supplies.

Response to the questionnaire survey showed the following:

- 187 responses were received from highway, sewerage and water supply authorities
- 53% considered highway runoff to be a source of pollution
- 47% did not consider highway runoff to be polluting (principally highway authorities)
- 80% of regulatory authorities had recorded significant changes in water quality adjacent to highway drainage outfalls, which were, or might be, attributable to the use or maintenance of roads
- 27% of water supply authorities have instigated measures to protect supplies as a result of pollution from a highway.

The distribution of pollutants in highway runoff is not regular or continuous. In particular there is the first flush phenomenon, which has been reported by a number of authors. At the start of a rainfall event the pollutants that have been accumulating on the road surface during the preceding period of dry weather (the antecedent dry period) are collected by the first movements of water on the road surface and through the drainage system. Strecker et al (1990) have suggested that the first 10% of the total discharge will be much more highly polluted than the remaining 90% of the runoff. This is an over simplification, for it has been shown that the actual nature of the storm can be a major determinant in the liberation and collection of road generated pollutants. The short duration, high intensity type of rainfall event typical of Britain in summer is generally accepted as being the most detrimental form of storm in pollution terms because of the abrasive quality of the high intensity rain and the limited volume of water that falls, thus mobilising a high percentage of pollutants and forming a concentrated effluent. This is in comparison to the longer duration winter storm that generates a greater total quantity of water, but tends to be less intense.

The questionnaire survey also indicated that at present there is very little provision on existing roads in Britain for the containment of pollution from highway runoff although the last half decade of major road building has seen a marked change in attitude. Of the highway authorities responding to the questionnaire survey, 30% said that some provision for treatment of highway runoff was made before it was discharged.

The survey revealed a disturbing lack of knowledge of the existence of policy or guidelines relating to discharge of highway runoff. Of the 187 replies received, only 71 declared that they were aware of any policy or guidelines relating to highway drainage. If the 100% return from NRA regions is ignored, it is clear that only 35% of the remaining authorities have an awareness of this issue.

4.3 HIGHWAY DRAINAGE METHODS AND DRAINAGE SYSTEM COMPONENTS

4.3.1 Highway drainage functions

Highway drainage systems generally fall into two categories: surface water drainage and sub-surface drainage. Some systems such as filter drains can satisfy both functions. The components of highway drainage systems fulfil one or more of the following functions:

- *collection* of water from the road surface or sub-surface
- *conveyance* of surface water to another point in the system
- *disposal* of collected surface water to watercourses or groundwater
- *storage* of surface water to reduce peak flows
- *coarse sediment removal* to prevent blockages
- *pollutant removal* to protect receiving waters.

For example, storage ponds are used primarily to attenuate peak flows but they can also be effective in the removal of sediment and associated pollutants. Under some conditions storage ponds can be used to isolate spilled pollutants and prevent contamination of receiving waters.

Collection of surface water

The following systems are in common use for the collection of surface water:

- kerbs and gullies (either trapped or un-trapped) including combined kerb and channel drainage units
- filter drains (previously known as french drains)
- surface water channels and ditches
- informal systems for dispersing water across or through a roadside verge
- precast channel or slot drains.

All the above require the surface water to be directed by design crossfalls to the edge of the pavement where it is collected.

Collection of sub-grade drainage

Sub-grade or sub-soil drains are usually designed to deal with small quantities of water which may percolate into the road construction, either through a poorly maintained road surface that would normally be impermeable, or by groundwater percolation. These small diameter porous or perforated drains, sometimes known as fin drains or narrow filter drains (DOT, 1991), are surrounded by a geotextile and/or a graded granular filtration system. Little, if any, allowance need be made in design calculations for the water yield from these drains except in the design of surface water channels (see Section 4.3.3) when up to 25% of the rainfall falling on the road should be allowed for as infiltration water into the sub-grade drainage system (HA37/88 DOT, 1988; 1991).

Conveyance

Once the water has accumulated in the collection system it needs to be conveyed by a suitable means to the disposal point. With the exception of informal verge systems and open highway drainage ditches, the predominant method employed for conveyance is a piped system.

Disposal

This is usually achieved by discharging the runoff to a receiving watercourse or a soakage system.

Table 4.1 summarises the major functions and water quality attributes (i.e. contribution to improvement of water quality) of drainage system components used in Britain.

Table 4.1 Functions and water quality attributes of highway drainage methods

Method	Primary function	Secondary functions	Water quality attributes	Report section
kerbs and gully pots	collection of surface water	sediment removal (gullies)	can remove pollutants associated with sediment but can also generate polluted liquor	4.3.2
filter drains	collection and conveyance of surface water	collection of sub-surface water	can remove pollutants associated with sediment but may be a risk to groundwater quality	4.3.3
surface water channels	collection of surface water	conveyance of surface water	none	4.3.4
porous surfacing	collection of surface water		unknown	4.3.5
precast channel or slot drains	collection of surface water	none	none	4.3.6
informal verge systems	collection of surface water	none	can remove pollutants associated with sediment	4.3.7
fin drains	collection of sub-surface water	none	associated geotextiles may remove pollutants associated with sediment	4.3.8
infiltration pavements	collection and disposal of surface water	sediment and pollutant removal	can remove pollutants associated with sediment and dissolved pollutants but may lead to increase in nutrient levels	4.3.9
catchpits and grit traps	sediment removal	pollutant removal	can remove pollutants associated with large sediments	4.3.10
oil separators	pollutant removal	none	can remove oil and other liquids that float on water	4.3.11
swales and ditches	conveyance of surface water	storage; sediment and pollutant removal; disposal	can remove suspended and possibly dissolved pollutants but may be a risk to groundwater quality if not sealed	4.3.12
infiltration basins	disposal of surface water	storage; sediment and pollutant removal	can remove suspended and possibly dissolved pollutants but may be a risk to groundwater quality	4.3.13
soakaways and infiltration trenches	disposal of surface water	storage; sediment and pollutant removal	can remove suspended and possibly dissolved pollutants but may be a risk to groundwater quality	4.3.14
detention tanks	storage of surface water	sediment and pollutant removal	can remove pollutants associated with sediments	4.3.15
storage ponds	storage of surface water	sediment and pollutant removal	can remove pollutants associated with sediments and provide some biological treatment	4.3.15
sedimentation tanks	sediment removal	pollutant removal	can remove pollutants associated with sediment and liquids not miscible with water	4.3.16
lagoons	pollutant removal	sediment removal; storage	can remove pollutants associated with sediment; vegetation can provide further treatment	4.3.17
wetlands	pollutant removal	storage	can remove and treat various pollutants	4.3.18

4.3.2 Kerbs and gully pots

Gullies are generally used in combination with kerbs. However, in rural areas and on lightly trafficked routes gullies are often installed without kerbs. The efficiency of the gully is considerably reduced where there is no kerb to generate the channel flow width.

The position and number of gullies installed in a highway will depend on the rainfall regime, the area being drained, the gradient of the carriageway and the type of surface. They are located at all low points, at street intersections and at intervals along a highway. The normal procedure is to provide sufficient numbers to prevent water from standing or crossing the carriageway (Bartlett, 1979). Spacing can be determined by road surface area (typically 200 m² per gully) and possibly the highway longitudinal gradient. More sophisticated methods are available, taking into account inlet configuration and flooded width (TRRL, 1969; 1973 & 1984), for paved areas (BSI, 1983) and for highways (HRS, 1989).

Gullies usually incorporate a "pot", a small sump permanently full of water, intended to trap sediment. Current advice (BSI, 1983) on the size of the gully pot required in a particular situation states that they "should normally be of adequate size, depending on the use of the area, the type of surface and the frequency of sweeping". Unfortunately, no definition of "adequate" is given or information on how it should be quantitatively related to the factors mentioned. The importance and impact of gully pot cleaning is discussed in Section 4.5.1.

A laboratory study in Britain (Karunaratne, 1992) reported solids-trapping efficiencies as shown in Table 4.2 (inflow up to 1.5l/s i.e. < 27mm/hr rainfall intensity). For a typical inflow solids grading, the overall total solids reduction would be 70-75%.

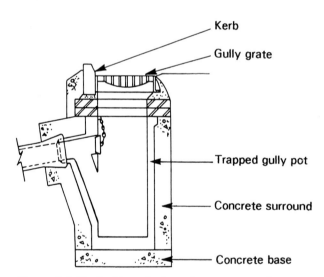

Figure 4.1 *Typical installation of a trapped gully with kerb*

Table 4.2 Trap efficiency of 450mm diameter BS gully pot   (Karunaratne 1992)

Sediment size (μm)	Efficiency
63 - 100	<= 35
100 - 150	<= 73
150 - 300	<= 95
> 300	> 95

Gully pots may have both a positive and a negative effect on the quality of highway runoff, although studies on gully pots in North America (where they are known as catchbasins) have been inconclusive. Aronson (1983) found gully pots quite effective for reduction of total solids

(60-97%) and associated pollution characterised by COD and BOD (10-56% and 54-88% respectively). Previously, however, Lager et al (1977) found that they did not remove appreciable quantities of pollutants and could in fact generate some. Similarly variable results were reported for German pots (Grottker, 1990a; 1990b). The extent to which the pot actually contributes to the total sediment load, due to material previously deposited in the bottom of the pot being mobilised by turbulence, is undetermined. In the laboratory study (Karunaratne, 1992), continuous erosion of sediment was only observed under the most onerous conditions (highest flow rate, deepest retained sediment bed and smallest particle size) and even then only amounted to a concentration of up to 10mg/l. However, continuous monitoring identified higher (150mg/l) transient peaks during the first 30 seconds of inflow. Field experience (Pratt et al, 1986; 1987) has shown the extent of erosion to be unpredictable. Pratt and Adams (1984), for example, found that the final suspended sediment concentration was less than half the initial sample concentration in only 18% of the catchment events studied whereas in 44% of the cases the final concentration was found to be higher than the initial value. The quality was always worse at flows in excess of 10l/s, which is equivalent to a rainfall intensity in excess of 180 mm/hr; this is a very rare event in the UK.

Detailed studies of gully pot liquor have identified its potential to act as a source of pollutant which can contribute to the deterioration of runoff quality (Fletcher et al, 1978; Mance & Harman, 1978). Pollutants can form between rainfall events, particularly when temperatures are warm, when the pot contents are liable to undergo significant biochemical changes including the build-up of anoxic conditions and anaerobic degradation of the bottom sediments. Digestion can in turn lead to the release of soluble organic compounds and increases in BOD, COD, ammonia and bacterial levels (Crabtree et al, 1991). Substantial quantities of heavy metals may also be flushed out by the first incoming runoff (Morrison et al, 1988).

In addition to the retention of larger sediments, gully pots can perform a limited oil removal function. The trapped gully is intended to do this by allowing immiscible material that is lighter than water to be retained above the free water surface from which discharge occurs. However, this function has the undesirable side-effect of increasing the tendency for anoxic degeneration of organic solids to take place.

In certain situations, gully chutes are used instead of pots. In a chute, sediment and other debris passes directly into the pipe system where it can be collected in downstream catchpits.

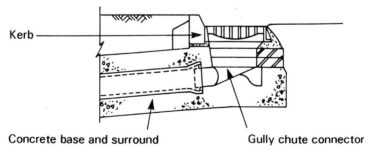

Figure 4.2 *Typical installation of a chute type gully with kerb*

For maximum efficiency, gullies are used in combination with kerbs. Consequently gullies are a moderate to high cost drainage system.

Gullies were found to be overwhelmingly the most popular means of collecting surface water runoff. Estimates suggest that they are used for at least 75% of all major roads. Approximately 70% of these are used in combination with kerbs.

4.3.3 Filter drains

Filter drains, previously known as french drains, are particularly suited for road drainage in cuttings. They have dual functions of collecting surface water runoff and controlling the groundwater level below the road. Narrow filter and fin drains are covered in Section 4.3.8.

Filter drain pipe sizes are calculated in a similar way to conventional storm water piped drainage systems. Care must be exercised in the detailing of the stone media (Spalding, 1970) and the pavement edge (DOT, 1989).

Filter drains have been widely used in a variety of road drainage applications. The main data available on the pollutant removal efficiency of this type of system was generated by a study commissioned by TRRL in 1980 on a $3172m^2$ area of the southbound carriageway of the M1 in Bedfordshire (Colwill et al, 1985). The treatment capabilities of three drainage elements were evaluated during the study: 55 metres of filter drain (see Figure 4.3), a 28m long lagoon and a sedimentation tank of $1.5m^3$ capacity. Mean annual removal efficiency data for a number of the parameters studied are shown in Table 4.3, which indicates that the filter drain system was very effective in the removal of solids and solid-associated pollutants. The effective lifetime of the filter drain was estimated to be in the region of ten years, after which time the voids between the bottom of the trench and drainage pipe were expected to be full of captured solid material. It was considered that once this occurred a substantial reduction in the treatment efficiency would ensue, and that removal and replacement of the granular material would be necessary. Equipment has been developed that allows filter material to be removed, cleaned and then replaced. Recent DOT policy has led to a decline in the use of filter drains, with the main reasons for this being given (DOT, 1991) as:

1. The cost of drain backfill.

2. The need for regular maintenance to prevent the growth of "grass kerbs" (build up of sediment and grass growth) that would inhibit inflow.

3. The cost of replacing filter material at approximately ten-year intervals.

4. Problems of stone scatter.

5. Softening of the pavement foundation and structure due to surface water infiltration from filter drains, particularly when the drain is clogged.

6. Risks of groundwater pollution.

With respect to 3., Colwill at al (1985) suggested that the working lifetime of a filter drain could be extended by increasing the depth of the trench below the invert of the drainage pipes.

Sub-surface drains are similar to filter or french drains except that the filter material is not extended to the surface but is capped by road construction materials or by soil. A porous geotextile sheet is sometimes used to envelop the pipe before the trench is backfilled with soil. The purpose of sub-surface drains is to prevent the inflow of water from the verge into the pavement and to collect and remove water from the pavement construction in order to maintain the optimum moisture content for strength, durability and stability.

The cost of installing filter drains is dependent on site location and geographical position. In general they are of moderate cost but in some circumstances, due to regional factors, they can be high cost (DOT, 1991; TRRL, 1987).

It is estimated that filter drains are used to collect surface water runoff from 25% of all major roads in Great Britain.

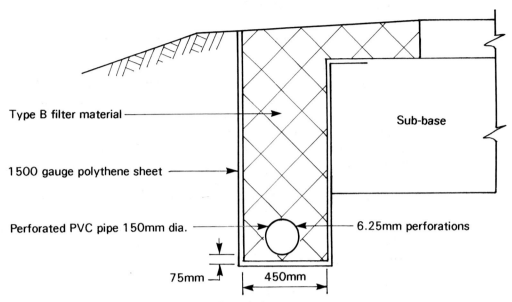

Figure 4.3 *Cross section of a typical filter drain*

Table 4.3 Mean annual removal efficiencies (%) for a filter drain, sedimentation tank and lagoon at an M1 experimental site (after Colwill et al, 1985)

Parameter	Sedimentation tank	Filter drain	Lagoon
Total suspended solids (TSS)	52	85	92
Total lead	40	83	90
Total zinc	47	81	76
Solid associated zinc	57	84	84
Dissolved zinc	15	56	62
Chemical oxygen demand (COD)	35	59	54
Oil	< 30*	70*	> 70*
Polynuclear aromatic hydrocarbons (PAH)	< 30*	70*	> 70*

Notes: * estimated value - not verified statistically

4.3.4 Road-edge surface water channels

Road-edge surface water channels are a more recent innovation. They combine the functions of surface water collection and conveyance and are usually longitudinal concrete channels of either triangular or rectangular cross-section (this latter form is now only permitted behind safety fencing on DOT schemes). The use of channels allows longer distances between outlets than conventional gully systems, and the positions of the outlets can be more easily chosen to suit the local environment. Water enters the channel directly from the road surface. Occasionally free overrun across the verge is employed but the verge must then be lower than the pavement and slope away from the highway.

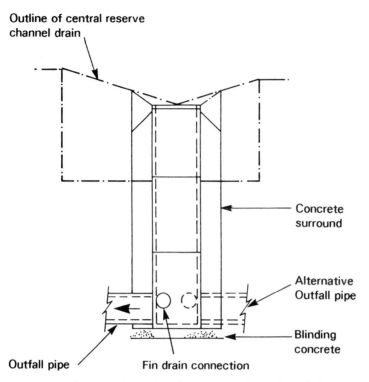

Figure 4.4 *Section view of surface water channel outfall*

The channels themselves are sized to ensure sufficient capacity under design storm conditions and to avoid encroachment of standing water onto the road (DOT, 1989a; 1991). Formulae suitable for incorporation in computer programs have been developed by HR Wallingford for the purposes of design. These are published in an Advice Note from the DOT (Design Guidance document HA37/88, 1989; Amendment No.1, 1991). Details of suitable forms of construction for these channels are included in the Highway Construction Details (DOT, 1991). Longitudinal gradients should be greater than 0.8%. The specification of the components of a surface water channel is critical and it is strongly recommended that specialist advice is retained for design.

The most widely used form of construction is a mechanical slip-forming or extrusion technique, using concrete. The choice of technique is largely a function of the physical dimensions.

Channels are easily maintained, if located strategically, by means of normal road sweepers and gully emptying machines.

They offer very little in terms of pollution control apart from their easy access for road sweeper maintenance. It may be possible to contain spillages by erecting temporary dams across the channels, provided that action can be taken with sufficient speed following an incident. This may be a very short time if rain is falling.

Surface water channels are high cost, particularly when parallel carrier drains are also required.

Channels are not widely used at present because their development is still in its infancy. Some 10% of motorways and dual carriageways, constructed during the last 4-5 years, are drained in this way compared with 5% of other trunk and county roads.

4.3.5 Porous surfacing

Porous surfacing consists of a porous macadam wearing course superimposed on a dense impermeable pavement construction. The benefits of porous surfacing are the reduction of splash and spray, reduced risk of aquaplaning, increased skid resistance and in particular reduced traffic noise (Watkins & Fiddes, 1984). The USA, France, Holland, Austria and Sweden have used porous macadam surfacing for both trafficked areas (public highways) and pedestrian areas (Diniz, 1976; Goforth et al, 1984; Hogland, 1990). Porous surfacing is being used by the DOT and other highway authorities where these benefits outweigh the additional costs.

Guidance on the application and design requirements of porous surfacing will be published in the *Design Manual for Roads and Bridges Volume 7* Section 1 Part 3 (DOT, 1994).

Porous surfacing probably exhibits a low pollution removal capacity.

Porous surfacing is expected to carry a 25% to 30% additional cost premium, and higher maintenance costs compared to conventional surfacing.

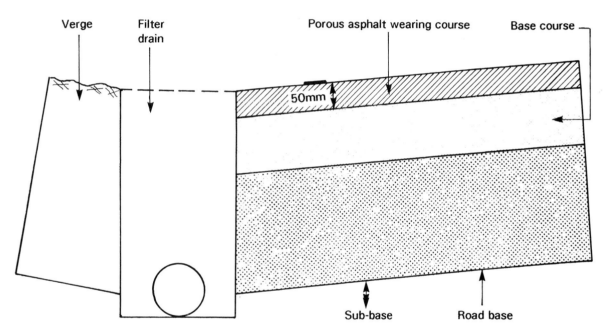

Figure 4.5 *Typical cross section of porous surfacing draining to filter drain*

4.3.6 Precast channel or slot drains

A relatively recent development is the combination kerb/channel system. In this system the kerb and channel are combined together and are capable, to a limited extent, of conveying runoff at a shallow depth over short distances. Discharge from the system is via conventional gully pots, but discharge points are fewer than if standard gullies were installed.

For minor roads, car parks and pedestrian areas there is a large range of channel systems available. Such products have holes or slots to allow drainage along their whole length. However, these are known to be difficult to maintain. More recently, in attempts to overcome blockage, open channel systems using continuous gratings have been used.

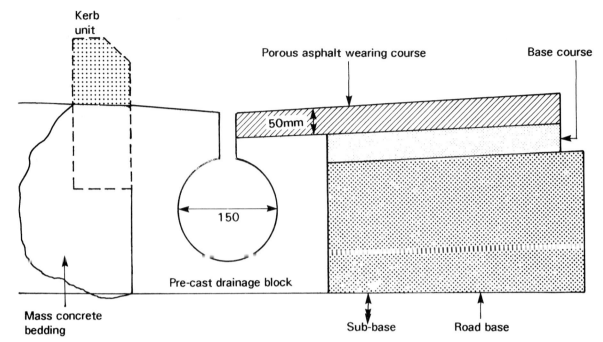

Figure 4.6 *Proposed application of slot drain for drainage of porous surfacing*

Such systems are of value to the highway drainage engineer in particular circumstances but do not have any definable benefit for pollutant removal over and above that of kerbs and gullies.

They have high installation and maintenance costs. Only about 3% of trunk roads use this system at present.

4.3.7 Informal verge systems

Primarily in rural areas, drainage of surface water, especially from single carriageway roads, is achieved by allowing the water to dissipate across roadside verges into ditches or adjacent land. This practice, termed "over the edge drainage" in DOT documents (HA39/89, 1989), was a feature of works under the 1854 Highways Act. Its continued use reflects the lack of any perceived need to change standards in subsequent years.

The main drawback to the employment of such systems is the tendency for vegetation to grow above the edge of pavement level, thus in effect forming a kerb that retains water on the road surface. A common alternative is the provision of grips or outlets built through a raised verge. These are usually open channels but can be piped. The latter method has road safety advantages but needs more regular maintenance to avoid excessive siltation (Pearson, 1990).

The advantage of such systems for water quality appears to be the natural vegetation and surface soils over which runoff is dissipated. Here pollutants are either entrapped and become combined with the soil particles or they are biologically degraded to non-polluting compounds. All such action is aerobic and therefore less likely to produce difficult secondary products.

There are still many highway authorities who actively discourage the replacement of such systems on the grounds of their low maintenance costs. The installation of "over the edge" systems on new highway works and major routes is negligible and it is probable that in time they will be replaced altogether by more sophisticated systems.

Survey results show that over-verge systems are still in evidence in Britain - usage is reported on 10% of major roads. Grips are thought to be even more prevalent, draining 20% of trunk/county roads.

4.3.8 Fin drains

In the early part of the century the merits of effective sub-soil drainage were commented on by Frost (1910), who observed: "There are just three things that are necessary to get a good roadbed and they are drainage, drainage and drainage". During the last three or four decades, however, the primary emphasis on pavement design has been on higher strength through density and stability of the roadbase. It has been widely assumed that the ability to remove water from within the construction is of less importance (Cedergreen, 1974). Indeed, the majority of British minor roads have no formal sub-soil drainage system. This is strange because even under moderate usage conditions it has been shown (Cedergreen, 1974) that if a road pavement is allowed to remain wet for 50% of its life, then that useful life will only be 10% of that of a perfectly drained section. Major roads conversely do have formal sub-grade drainage as specified in DOT Design Guidance Notes. Recent work by Hillary et al (FHA, 1992) has shown that it can be beneficial to install sub-soil drainage, in the form of edge drains, on existing roads.

Recently, sub-surface drainage systems have included vertical fin drains. Doubt has been expressed (Pearson, 1990) over the long-term performance of these items and their practicality from a maintenance standpoint. The design for satisfactory installation, as part of new works has caused difficulties and is an area that has only recently received limited study. The size of the pore spaces in the geotextile used for the fin drains is critical and is decided by the O_{90} characteristics of the chosen roadbase/sub-base (that is a pore size just larger than 90% of the pores in the material). The flow design is based on a notional 25% percolation rate for rainfall on the road surface.

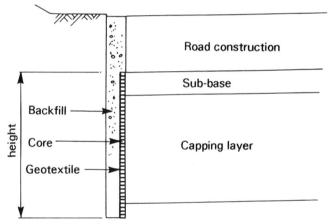

Figure 4.7 *Fin drain in edge of pavement*

The Highway Construction Details (DOT, 1991) also allow for narrow filter drains to be installed. These are much less costly to install than fin drains but are subject to the same design uncertainties and, because of their unsophisticated construction, doubts have been expressed about their performance (Corbet for TRL, 1990).

The ability of fin drains/narrow filter drains to contribute to the removal of pollutants from highway runoff is limited because they are designed to deal solely with sub-grade water.

The direct cost of fin drains is less than conventional, wide filter drains (see Section 4.3.3) but as they do not have the capacity to act as drains for surface runoff, there is an additional cost in providing a separate system for that function. It is reported that fin drains and a separate system for runoff cost 10 to 15% more than filter drains for similar situations.

The maintenance of these drains is difficult because of their small size. Any malfunction would generally require replacement.

4.3.9 Infiltration pavements

Infiltration pavements, as the name implies, allow water to move vertically through the construction to be collected at formation level or to dissipate into the subsoil. This form of construction is not permitted for DOT schemes and is only rarely used in the United Kingdom and then only for lightly trafficked roads or car parks. Infiltration pavements are capable of attenuating the runoff by acting as storage devices and hence may be used as a flow control option. Although many of the early experiments with this construction were carried out in Britain (Brown, 1973; Szatkowski & Brown, 1977), very little operating experience is available. Figure 4.8 shows a range of different types of construction.

There are many examples of grass concrete infiltration pavements in other countries (Pratt, 1992), most commonly off the public highway particularly where landscaping considerations are important. Examples can be found in residential areas and car parks. The wider application of such construction has been called for (Day et al, 1981). Interlocking concrete blocks have been more widely used but it is believed that Japan is the only country to have utilised them on public highways for stormwater management purposes (Fujita, 1984; Suda et al, 1988).

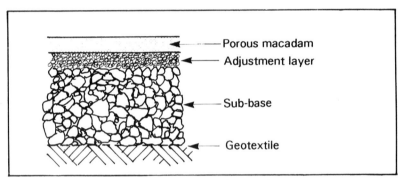

Permeable pavement with porous macadam surface

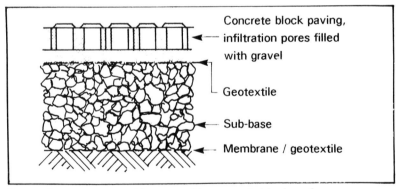

Permeable pavement with permeable concrete block surface

Figure 4.8 *Typical cross sections of infiltration pavements*

Niemczynowicz (1989), quoting Hogland, reports that some $7 \times 10^6 m^2$ of infiltration pavement has already been constructed in Sweden.

Sources of current design advice are limited (Pratt & Hogland, 1990; Leonard & Sherriff, 1992), although in principle the major requirement for plane infiltration is that the rate of infiltration of the surface (permeable surfacing, macadam or concrete cellular blocks) should be at least equal to the design rainfall intensity. Fully permeable roads with no underdrainage are not recommended for highways since the structural integrity of the pavement may be compromised (Degroot, 1982).

Infiltration pavements have the potential to reduce flow and also enhance its quality. The principal mechanisms of pollutant retention are thought to be sedimentation, filtration and chemical adsorption onto materials within the pavement construction (Pratt, 1989). If infiltration pavements are therefore designed so that the percolating water can be collected in underdrains, the partially treated water may be discharged for further treatment or diverted to a watercourse.

Considerable research into infiltration pavements has been carried out in Europe. Significant removal of heavy metals has been recorded, although removal of phosphorus forms appears to be very variable and nitrites/nitrates can be added to the effluent due to leaching from the soil. It has been found that a pavement with a blast furnace slag sub-base is effective in storing runoff but in the short term suspended solids in the runoff increase. In the longer term, however, suspended solids concentrations can be reduced to as low as 20mg/l and are less than 50mg/l in most cases. Studies indicate that the effluent water quality is dependent on the sub-base material used (Day et al, 1981; Pratt et al, 1989). Concern exists over the long-term retention of pollutant spillages and the subsequent protracted release with each rainfall event.

Although the construction cost of an infiltration pavement is greater than the cost of a conventional pavement, costs savings may be made because of the reduced drainage system required. Maintenance costs for the surfacing will be higher but lower drainage maintenance costs will offset this. Factors other than cost will often influence the choice of infiltration pavements.

4.3.10 Catchpits, grit traps and manholes

Catchpits or manholes are constructed at drain junctions, changes of gradient and certain other locations to allow access for maintenance of the drain. The size of the chamber is governed by the depth of pipes, the requirement for man entry and the size of the incoming and outgoing pipes, as there must be sufficient space to allow connection to the chamber wall. In catchpits and grit traps, provision is made for sediment to be collected, whereas in a manhole runoff and associated sediment is usually channelled through the chamber.

Although catchpits and grit traps are very widely used on British highways, no extensive studies have been carried out to measure their efficiency in removing contaminants from runoff. Some preliminary data from Milton Keynes and France (Ellis, 1991b) suggest, however, that they may remove up to 25% of sediment, 10% of metals and nutrients and 30-40% of oil. Manholes are assumed to have no pollutant-reducing capability.

Contaminant removal efficiency is limited by inherent design factors:

1. The limited amount of wet storage provided in the chamber(s), which is equivalent to between 0.7mm and 0.8mm of runoff per impervious hectare. This is less than half the storage normally recommended for removal of the first flush contaminant load. Fine grained particulate contaminants and soluble species are even less likely to be removed to any significant degree.

2. Short detention times - from a few minutes to 30-40 minutes.

3. Sediments and contaminants deposited during small storm events may be re-suspended and removed during the next large event.

4. Routine maintenance needs are high because of the first two items. Ellis (1991b) calculates, however, that a storage volume of $1.5m^3$ per impervious hectare should limit sediment removal requirements to once per annum.

Maintenance consists of the removal of sediment from catchpits and grit traps and the cleaning of the drain lengths that connect to the chamber.

Catchpits and grit traps are expensive. Economies can be made by careful location at strategic points and by due consideration of the design capacities required.

4.3.11 Oil separators

The purpose of an oil separator is to capture oil from a liquid and to hold it separated from the transporting fluid until it can be dealt with. The operation of oil separators (also known as oil interceptors or oil traps) is dependent upon the oil product being immiscible in and lighter than water.

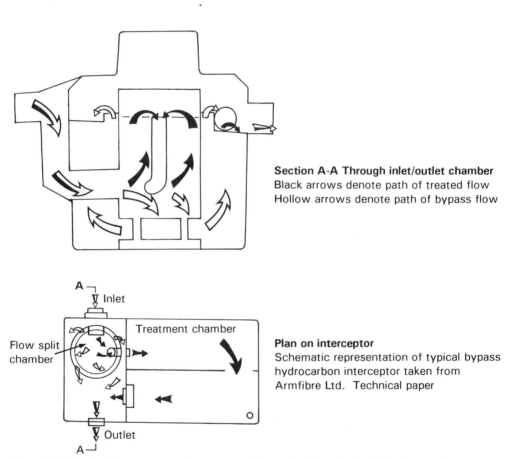

Section A-A Through inlet/outlet chamber
Black arrows denote path of treated flow
Hollow arrows denote path of bypass flow

Plan on interceptor
Schematic representation of typical bypass hydrocarbon interceptor taken from Armfibre Ltd. Technical paper

Figure 4.9 *Typical bypass type oil separator (from Armfibre Ltd technical paper)*

The normal method of separation is by displacement using baffles, dip pipes or combinations of the two. Bypass type separators, which work on a modification of the same principal, are gaining favour. The primary difference between these and the normal forms is that a system is incorporated that limits the throughput of effluent receiving full treatment. This recognises the greater pollutant load carried by the first flush of a runoff event. Typically full treatment is only given to 10% of the design maximum runoff. Flows in excess of this allowance will theoretically contain a much lower pollution load and be subject to a higher dilution ratio. These flows are allowed to bypass the treatment system. Consequently the physical dimensions of the trap can be smaller than for a conventional separator. The design has the added advantage of avoiding the situation where excessive turbulence, created by high flows, can flush out accumulated quantities of oil, as has been noted with conventional type separators.

Only recently have attempts been made to come to terms with some of the more difficult scenarios of oil products in water. The turbulence created by high capacity flows tends to cause break-up of some of the oil fractions into minute globules, which are carried within the body of the flow rather than on the free water surface. Newer systems have been developed, including tilted plate separators and most recently coalescing filters, which are incorporated in a new family of separators covered by draft European Standard prEN 858-1 (1992). These devices, developed for the petro-chemical and food processing industries, are capable of removing the oil load with greater efficiencies than more traditional installations. Costs, however, are also increased.

All oil separators tend to capture sediments at their base. This is not always allowed for either in the original design or in the maintenance scheduling and can have far-reaching effects. The most common design criteria are for the provision of a specified storage capacity, which in most cases will reflect a minimum time of retention at design peak flow. In some cases, the ability to hold the entire capacity of a tanker (normally $20m^3$) is also required so as to provide containment of accidental spillage. One further aspect of oil separator design, which tends to be overlooked, is that they require regular maintenance involving mechanical emptying. Thus they must be accessible to wheeled vehicles large enough to carry out such maintenance operations.

The inclusion of oil separators in a drainage system will normally be as a response to standards required by the pollution control authority. They are expensive items and therefore economic design must locate them in the minimum number of strategic locations.

Although the removal of petroleum products from road runoff has been discussed for very many years, the responses to the questionnaire survey conducted for this report indicate that only about 10% of highway runoff discharges are currently receiving any direct treatment for the removal of oil products.

4.3.12 Swales

Swales have only recently appeared in European highway drainage. In North America they have been a major element of drainage system design for some time. In form they are shallow, grass-lined depressions, which can offer some degree of infiltration. They are used for the conveyance, storage and infiltration of runoff, in much the same way as roadside ditches in Europe.

Runoff from highways is either discharged down the side slopes or conveyed from drains into the swales. Swales can also be used to convey runoff to a central infiltration basin or can have an overflow into the piped drainage system.

A tentative design method for British conditions has been proposed (Leonard & Sherriff, 1992). Since the main form of maintenance required is grass/vegetation management, side slopes should be designed to allow easy access for such maintenance by mechanical plant.

As a method of runoff conveyance, storage and infiltration, some degree of water quality improvement is to be expected with the use of swales due to particle sedimentation and bio-filtration. Ellis has reviewed the factors that influence the treatment capability of swales (e.g. design infiltration rate, type of grass, maintenance practice) and how these may be optimised (Ellis, 1992a; Ellis 1991b). However, current British design approaches (CIRIA, 1992; Ellis, 1991b) do not consider the potential for water quality improvement. The only information available on the treatment performance of swales originates from American studies (see review in Ellis, 1991b) which indicate significant removals of solids, hydrocarbons, bacteria, heavy metals, total nitrogen and total phosphorous. Actual removal values, however, were highly variable. This variability of performance is to be expected given the inherent diversity of operating conditions associated with swales or grassed channels. An indication of the level of treatment that might be achieved is shown in Table 4.4, which presents results from an American study of two freeway interchange sites (Yousef et al, 1987).

The general performance of roadside ditches in Europe is very similar to swales and it is reasonable to expect that ditches will produce equivalent results. Evidence of this has been reported in Britain by Colwill et al (1985), but no complete investigation or evaluation has been carried out. The use of such techniques must be conditional upon the suitability of the underlying geology, especially in terms of any potential for contamination of aquifers.

The installation of swales is likely to be in the low to medium cost range, but they do require the provision of land suitable for both the construction and future maintenance requirements and so perhaps are more appropriate in a rural rather than urban context. However, for new residential developments where runoff attenuation or on-site storage is required, swales can be incorporated into soft elements of design.

Table 4.4 Characteristics and removal rates for grass swales (after Yousef et al, 1987)

		Maitland Interchange				Epcot Interchange			
Swale Length (m)		53				90-170			
Average cross-sectional area (m^2)		0.014 - 0.63				0.056 - 0.071			
Hydraulic depth (m)		0.017 - 0.038				0.040			
Input volumes (m^3)		8.14 - 40.9				30.8 - 57.7			
Inflow rates (m^3/m^2.hr)		0.036 - 0.154				0.053 - 0.094			
Infiltration rates (m^3/m^2.hr)		0.024 - 0.088				0.007 - 0.034			
Average mass removal rates (mg/m^2.hr) [% reductions]	Pb	1.14 [56%]				2.61 [76%]			
	Zn	1.85 [93%]				5.76 [77%]			
	Cu	0.42 [70%]				0.6 [49%]			
	Cd	-				0.26 [63%]			
	P_{total}	17.1 [63%]				13.96 [42%]			
	N_{total}	85.3 [51%]				44.8 [41%]			
Average concentrations with distance along swale (mg/l)	Length (m)	Pb	Zn	P_{total}	N_{total}	Pb	Zn	P_{total}	N_{total}
	0	9	22	415	2049	67	140	599	2273
	30	8	9	367	2395	43	103	586	2456
	53	5	3	310	1817				
	90					41	77	558	2127
	170					29	53	580	2435

4.3.13 Infiltration basins

This section covers two different types of construction.

Infiltration basins are more commonly used in North America. They consist of a dry retention basin where runoff is stored until it infiltrates into the ground. There is no positive outfall; instead an emergency overflow is normally installed at a level determined by the sensitivity of the surroundings to flooding. The design requirements are similar to those of balancing ponds (Maskell & Sherriff, 1992). The use of infiltration basins is increasing in Britain.

Filtration basins are more commonly used in mainland Europe for treatment of runoff. They consist of sand and/or gravel filter beds with an underdrainage system to remove the filtered runoff. No British-based performance data are available but a number of studies have been undertaken in France and America (Ellis, 1991b). A typical design used in French studies is given in Figure 4.10, and corresponding removal rates in Table 4.5.

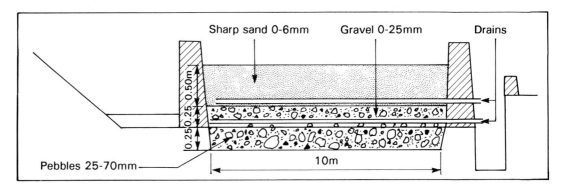

Figure 4.10 *A French filtration basin for treatment of motorway runoff (after Ellis, 1991b)*

The data from these studies indicate that, although performance can be effective, considerable reservations exist about the long-term performance of such systems. The French studies indicated that a silt crust tended to form over sand surfaces, effectively sealing the basin and causing surface ponding as well as generating anaerobic conditions. US studies showed that other reasons for failure were unsuitable soils, underdrainage and poor maintenance, and it was considered that infiltration basins have one of the highest failure rates for urban runoff management techniques in the US.

Nevertheless, the concept appears to be favoured in Germany where guidelines have recently been published by the German Road and Traffic Research Association (Lange, 1990).

The costs of construction, maintenance and land to install infiltration basins are all high in the United Kingdom. This probably accounts for the limited use made of this technique.

Table 4.5 Removal efficiencies for French motorway infiltration basins.
(after Cathelain et al, 1981; Ruperd, 1987; Rancher and Ruperd, 1983)

Site	Total area drained (ha)	Runoff co-efficient	Mean percentage removal					
			BOD	COD	TSS	Pb_{total}	Zn_{total}	Oil
St. Quentin A26	33.0	0.17	33 (0-73)	35 (0-75)	34 (0-73)	41 (1-67)	45 (0-48)	53 (7-86)
Lille A26	2.5	0.63		45	80	63	53	41
Bordeaux A10	10.7	0.63		17	33	44	26	

(Ranges, where known, are given in brackets)

4.3.14 Soakaways and infiltration trenches

A soakaway consists of a pit provided with a roof slab and an open jointed (i.e. pervious) base and sides, which may be dry wall, concrete or geotextile lined, or constructed from precast concrete rings. Frequently the pit is filled with stone or rubble.

Infiltration trenches take the form of filter drains but normally with the pipe omitted from the base. They perform a similar function in allowing water to percolate into the ground, although they are generally longer, narrower and shallower and also often covered by grass (see Figure 4.12). This configuration maximises the surface area for infiltration and hence infiltration trenches tend to be more efficient.

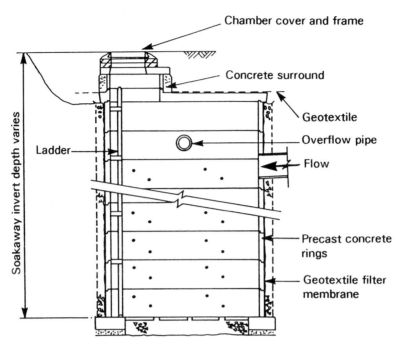

Figure 4.11 *Section through soakaway*

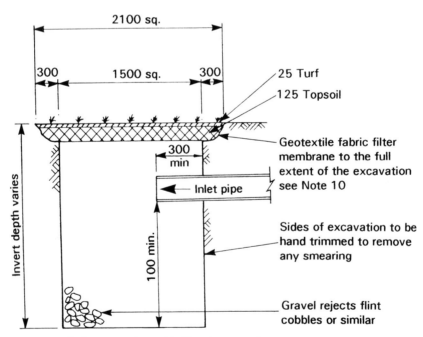

Figure 4.12 *Cross section of infiltration trench*

Both systems operate on the principle of providing storage for the stormwater until it can percolate into the surrounding ground. They are most effective on pervious subsoils such as gravel, sand, chalk or fissured rock, with low water tables (BSI, 1985). Soakaways without rubble infilling (underground cisterns) may also be used and may allow the opportunity for some form of water reuse (Fewkes & Jay, 1992).

There are many examples, in Britain and overseas, of soakaways serving one or two households (yards and roofs), or connected to single or multiple gullies for disposal of highway drainage (Bartlett, 1979). It is also common practice in highway drainage to link together multiple soakaways, which allows the storage and infiltration capability of all the soakaways to be used prior to overflow. However, this practice can lead to the downstream system being damaged by oil or other pollutants (Beale, 1992).

The most up-to-date guidance on soakaway design is based on the concept of minimum infiltration rate, which requires that the soakaway is half empty within 24 hours following rainfall (BRE, 1991). Further work of a more empirical nature has been conducted into the adequate sizing of soakaways by Oxfordshire County Council (Welham and Hunt, 1992). Numerical modelling of soakaway performance is the subject of ongoing study (Watkins, 1991; Payne & Watkins, 1992) and CIRIA research project RP448, when completed, will provide a manual for the design of infiltration systems.

Soakage can have pollution control benefits but these would result from the filtration by any infill that is employed, or by the strata immediately around the disposal point. Pollutants so captured might then be released at a slower rate by further water movement through the strata. The efficiency of soakage depends upon the permeability or porosity of the receiving stratum. The better this stratum is for soakage, the greater is the probability that it is also an aquifer.

Since the publication of the National Rivers Authority groundwater protection guidelines (NRA, 1992), the continuing application of soakaways for highway drainage disposal must be in doubt. More onerous conditions now exist to control soakage into aquifers and the guidelines indicate non-acceptability or a "presumption against" soakaways draining highways in groundwater source protection zones. It is clear that consideration of design and location will need special attention and early consultation with the pollution control authority is going to be essential.

Soakaways are used widely in the United Kingdom because they are believed to be a cheap alternative to positive discharge systems in certain circumstances. However, the cost of long-term maintenance is often overlooked, particularly as soakage is more prone to failure through the result of poor maintenance than positive discharge systems. If a soakaway fails, the cost of refurbishment is high. The capital cost is slightly higher than a manhole or catchpit but for many soakaways the cost of a proper maintenance regime is much higher. However, carefully designed and constructed soakaways should not have high maintenance costs.

4.3.15 Storage ponds and detention tanks

Although employed principally as flow attenuation and control devices, surface storage ponds will, in most cases, provide some degree of water quality improvement. Flow detention will lead to settling out of the particles and associated contaminant loads. In addition, some bacterial die-off and soluble contaminant reduction (e.g. metals, nutrients) may occur. The extent of the treatment achieved will depend on the type of storage pond (e.g. dry or wet), the mean flow detention time, and the pond design (e.g. the magnitude of hydraulic short-circuiting, the inclusion of wetland plants). In surface storage ponds, the storage is achieved either by a marsh area where surface ponding can occur or by formation of a pond with containment banks (Figure 4.13).

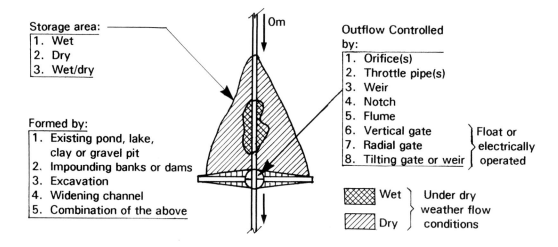

The onstream storage pond

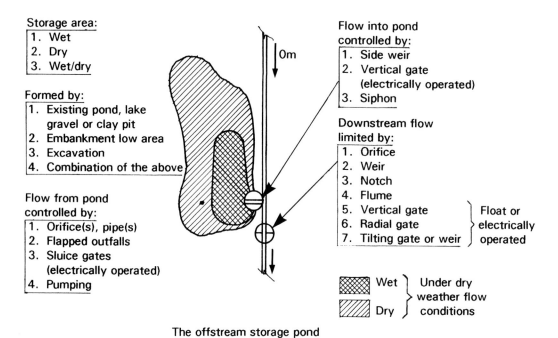

The offstream storage pond

Figure 4.13 *Typical on-stream and off-stream storage ponds*

The different types of storage pond have been classified by Hall et al (1993) as shown in Table 4.6.

Table 4.6 Definition of surface storage pond types   (after Maskell & Sherriff, 1992)

Type of Pond	Description
On-stream	Dry weather flow passes through the storage area
Off-stream	Dry weather flow bypasses the storage area
Dry	Storage area is free of water under dry weather flow conditions
Wet	Storage area contains water under dry weather flow conditions
Wet/dry	Part of the storage area contains water and part is free from water under dry weather flow conditions

Table 4.7 Detention basin removal efficiencies

Location	No. of storms	Total solids	P_{total}	N_{total}	Pb_{total}	Zn_{total}	Depth (m)	VB/VR	SA/SD (%)	Average detention time (days)	Reference
US EPA NURP Studies											
Lansing	65*	⎫									
Ann Arbor	16	⎪									
Long Island	8*	⎬ 52 (5-91)	31(3-79)	18(0-60)	62(9-95)	23(3-71)	3.5(1.5-5.0)	3.3(0.05-10.7)	1.08(0.01-2.85)	25(0.37-96)	Bascombe et al, 1988
Washington DC	32	⎪									
Chicago	23	⎭									
Orlando, Florida	18	58	42	40	62	49	6.0	1.0	1.4	7.3	Winters & Gidley, 1980; Yousef et al, 1987
Canberra, Australia	-	(62-94)	(30-85)	(38-76)	(58-87)	-	3.4(2.5-4.7)	0.14(0.01-0.53)	6.5(0.001-14.6)	-	Hamilton & Harrison, 1991
London, UK	24*	56(46-84)	21(10-58)	-	52(40-66)	38(8-56)	3.1	0.6	0.4	13	
Viborg, Denmark	4	(55-68)	-	-	-	-	5.8	(3.2-6.1)	-	-	Mance, 1981

Figures given are % reductions in mass load with ranges (where known) given in brackets.   * Some negative removal efficiencies were recorded.

VB/VR = Basin volume/mean storm runoff volume   SA/SD = Basin surface area/surface drainage area

Thus a large degree of variability exists as indicated in Table 4.7, which presents percentage reductions in the overall mass load that have been reported for a number of wet storage ponds in the USA, Europe and Australia, some of which received highway runoff.

In the questionnaire survey a small but significant number of replies indicated that some form of flow storage or containment structure is used as a deliberate method of pollution treatment within the highway drainage system or at the point of discharge.

Surface ponds, which are becoming more widely used for the control of highway drainage, are most appropriate to store runoff provided that it is not contaminated with foul sewage.

For wet ponds, Ellis (1991a) reported that the maximum pollutant removal capacity appears to be improved when the ratio of the pond surface area to the drained road surface area is in excess of 2 to 3%. This broadly agrees with the results obtained from a British research study of a lagoon treatment system on the M1 Motorway (Colwill et al, 1985). For dry ponds, a summary of results for a series of detention basins in Germany, studied during the last 13 years, indicated a disappointingly poor level of performance (Table 4.8), which was ascribed to the effect of re-entrainment of settled solids during high flows. The highest average removal rates were obtained for a pond fitted with a two-stage sediment/oil trap device at the inlet.

Occasionally on-line detention tank storage is used. In simple form this can be achieved by oversized pipes whereas larger installations may require a multi-barrelled tank. The outflow from these tanks is normally controlled by a fixed device such as an orifice, vortex regulator or throttle pipe. Off-line tanks, which incorporate inlet and outlet flow controls, are an alternative form commonly used in Britain. The inlet control comes into operation if the downstream sewer is not capable of transporting the available stormwater. The stored water is then returned to the system either by gravity or by pumping. Sedimentation may occur, especially if the flow is highly laden with sediment.

A detailed review of current practice for siting and design of detention tanks has been produced (Knott & Taylor, 1985). Advice on designing for maintenance has also been published (Sewers & Water Mains Committee, 1991). The influence of tank design on sediment deposition is currently the subject of investigation (Saul & Ellis, 1992).

Storage ponds are expensive because they require large areas of land and probably extensive earthworks. This cost can be reduced if they can be sited in surplus land (land that is no longer of any use to the owner), or landscaping areas. Maintenance costs are not high provided the design allows ready access for mechanical plant. Generally wet ponds are cheaper than dry ponds.

4.3.16 Sedimentation tanks

The 1980 TRRL study of the M1 in Bedfordshire, referred to in Section 4.3.3, included evaluation of the contaminant removal efficiency of a sedimentation tank with a nominal holding volume of 1500 litres, width 0.72m and length 3.49m. Its effectiveness in reducing particulate loadings showed significant seasonal variation (Colwill et al, 1985), averaging 35% in winter to 80% in summer. Further details are given in Table 4.3. Other studies of similar systems have been carried out in France. Ellis (1991b) reported that the limited data from these studies indicated similar removal efficiencies for COD, TSS (Total Suspended Solids) and total lead to those found in the TRRL study.

The TRRL study included estimates of the desludging frequency necessary to maintain performance, and derived a figure of 7-10 years. Based on the data from this work, Ellis (1991b) calculated the required dimensions of a similar tank for treating runoff from an area of typical motorway 1km long and 15m wide. Assuming 100% theoretical solids removal efficiency and a detention time of 30 minutes, they estimated that a tank of depth 0.57m, length 21.97m and width 4.55m would be necessary. They concluded that, given the lower removal efficiencies evident for soluble and organic contaminants, such a tank would not offer sufficient water quality improvement if used alone.

Taken in isolation sedimentation tanks are expensive to construct. However, as part of a comprehensive design, their cost would be offset by savings on catchpits and gully pots. The maintenance cost is high but, because the number of installations to be visited is reduced and there would be a longer period between visits resulting from the larger capacity, this would be lower than for an equivalent gully pot system.

4.3.17 Lagoons

Lagoons may be regarded as differing from sedimentation tanks only in the type of construction. Lagoons normally take the form of earth basins or roadside excavations, whereas sedimentation tanks are constructed using conventional building materials. The difference in the construction makes for an important additional difference in as much as lagoons can be planted with aquatic vegetation to provide treatment for polluted runoff.

Very few studies have been carried out to assess the efficiency of lagoons in the treatment of highway drainage water. An early study by Borch-Jensen (1978) evaluated the treatment efficiency of a 40m^3 lagoon receiving runoff from an area of Danish motorway containing 1.8 ha of carriageway and 2.1 ha of verges. Runoff quality was examined in terms of TSS, COD, Pb, Zn, N_{total} and P_{total}. Good removal rates were found for all six determinants, with averages over three events of 98, 69 and 95% for TSS, COD and Pb, respectively. Reductions were attributed to both sedimentation and adsorption of contaminants by plants in the lagoon.

A more extensive study of three lagoon-size detention basins in Germany has been reported by Stotz (1987, 1990). The pollutant removal efficiencies given, appear generally less significant than those reported by Borch-Jensen, which may reflect the apparent absence of vegetation at the German sites. At one site, at the A8/B10 Ulm West intersection, studies were also carried out to compare removal efficiencies of the lagoon operating in both dry and wet modes. Measurable improvements were noted in removal efficiencies of TSS, COD, P_{total}, Cd and other metals when the lagoon was operating in the wet mode (Table 4.8).

The 1980 TRRL M1 study, previously referred to, included evaluation of a lagoon 28m long and 2m wide, of approximately trapezoidal cross section with a maximum depth of 0.3m (Figure 4.14). The holding capacity was generally maintained between 3 and 5m^3 and could be altered by varying the height of the outlet pipe. In general, treatment efficiencies were extremely good (Table 4.3), and the very high solids removals were relatively unaffected by variations in flow rate, particle size distribution and input solids concentrations. Considerable water loss from the system was noted during the study, which may have contributed to high apparent treatment levels, but over-design was considered to be the main factor. Overall, the TRRL study found that treatment efficiencies for most materials in the three systems studied followed the order:

$$\text{lagoon} \geq \text{filter drain} > \text{sedimentation tank}$$

The design and maintenance requirements of lagoons are discussed in the publications dealing with storage ponds and sedimentation tanks (i.e. Knott & Taylor, 1985; Sewers & Water Mains Committee, 1991; Saul & Ellis, 1992).

Lagoons are expensive because they require large areas of land and extensive earthworks that will include some form of waterproof membrane and planting. This cost can be reduced if they can be sited in surplus land or landscaping areas. Maintenance costs are not high if the design allows ready access for mechanical plant. Adequately designed lagoons can be cheaper than dry ponds because of their self-sustaining ecology.

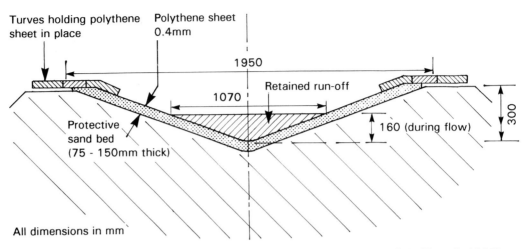

Figure 4.14 *Cross section of an experimental treatment lagoon (from Colwill et al, 1985)*

Table 4.8 Percentage removal efficiencies for dry type oil removal/detention basins in Germany (after Stotz, 1987; 1990)

	Location			
	Ulm - West		Pleidelsheim	Obereisesheim
	Dry mode	Wet mode		
Catchment area (ha)	25		1.3	2.52
% impermeable	86		100	40
Average daily traffic flow	41 000		47 000	52 100
Volume of detention basin (m³)	1090	1000	194	500
Oil Separator:				
Surface area (m²)	40		107	10
Surface loading (m/h)	9		9	18
Contaminant removal efficiencies (%)				
SS	45	54	85	50
COD	18	39	63	26
NH_4^+-N	10	-72	36	16
P_{total}	3	12	32	9
Cd	14	60	63	28
Cv	-60	7	66	33
Cu	-13	17	73	26
Pb	33	52	79	39
Zn	24	29	50	37
Fe	24	38	74	45
Motor fuel	16	33	80	29
Mineral oil	17	33	80	29

4.3.18 Wetlands

The term "wetlands" is generic since it includes natural and constructed land areas which are saturated for most of the year, through which runoff flow passes in a vertical and/or horizontal direction. Wetlands include reed beds, reed marshes and non species specific vegetative systems. Such systems have been used previously for the treatment of municipal, industrial and agricultural effluent (Cooper and Findlater, 1990), including urban stormwater, but very little has been reported on their use for highway runoff quality control. Only a small number of replies to the questionnaire indicated that highway runoff currently receives either wetland or reed bed treatment.

According to Ellis (1991b), there is general consensus on the principal treatment mechanisms in wetland systems and these are: bio-filtration, sedimentation, adsorption, biological uptake and physio-chemical interactions. However, there appears to be little appreciation of the relative importance of these mechanisms and how they are influenced by design variables. A fuller description of the treatment mechanisms can be found in Ellis (1991b).

The very limited number of studies concerned with the treatment capabilities of wetland systems for highway runoff indicate that a moderate-to-high degree of water quality improvement can be achieved. For a reed bed treatment system (e.g. Figure 4.15), Ellis (1991b) reported high pollutant removal efficiencies for an experimental system using *Typha latifolia* as the reed type; see Table 4.9. Verniers and Loze (1985) described the performance of a wetland basin treating stormwater runoff in the Paris area employing *Scirpus* and *Juncus spp* plants. Average removal efficiencies of 76, 55 and 17% were found for TSS, BOD and COD, respectively. Nutrient removal was between 8 and 30%.

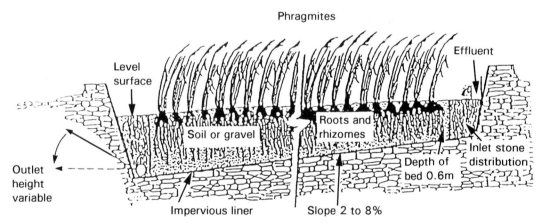

Figure 4.15 *Typical arrangement for a reed bed treatment system*

Table 4.9 Performance of experimental reed bed systems   (Ellis, 1991b)

Pollutant	Inflow	Outflow	% efficiency
BOD	200 (80-490)	21 (4.5-61)	89.5
COD	379 (125-850)	64 (24-118)	83.1
TSS	239 (30-630)	5.7 (1-12)	97.6
P_{total}	6.5 (1.9-15.4)	3.3 (0.5-10.6)	49.2
N_{nh}	18 (17-99)	26 (9.8-39)	45.8
Total coliforms	107×10^3 ($20-200 \times 10^3$)	36 (10-120)	99.9
Faecal coliforms	79×10^3 ($5-140 \times 10^3$)	21 (5-80)	99.9
Enterococci	18×10^3 ($4-40 \times 10^3$)	7 (0-40)	99.9

Average data values are given in mg/l. Ranges of values are given in brackets.
Bacterial data are given in colony forming units (CFU) per ml ; n = 84.

Martin and Smoot (1988) have presented results on the performance of an artificial wetland system receiving stormwater runoff from a highway in Orlando, Florida. Substantial pollutant removal rates of between 55 to 75% were recorded for TSS, metals and ammonia, while much lower removal efficiencies were found for phosphorous, nitrogen species, organic carbon, BOD and COD. There is growing interest in combining wetland systems with other treatment or flow control stages (Hall et al, 1993; Ellis, 1991; 1992) to provide a greater treatment capability. For example, Ellis has proposed a system based on a sequence of different macrophytic species (Ellis, 1991b). Meyer (1985) has reported on a study of a combined detention basin and artificial wetland for the treatment of highway runoff, in which high removal rates for TSS, metals, hydrocarbons, phosphates and nitrogen were achieved.

At present the basis for design of wetland systems is, at best, tentative, although useful empirical information is available in the literature. In particular, Ellis (1991b) has identified various sources of information that consider the vegetation sequences and basin geometry required to maximise biological uptake mechanisms (e.g. Hall et al, 1993; Ellis, 1990a; 1991; Cooper and Findlater, 1990).

The costs of construction and maintenance are similar to those for lagoons. The principal difference is in the costs of vegetation systems and their regular maintenance.

4.4 CONTROL OF PROBLEM POLLUTANTS

All established and validated anti-pollution techniques only treat those pollutants that can be physically separated from the runoff water. This is usually achieved by the provision of facilities to allow the settling out of sediments or the physical entrapment of immiscible pollutants. There will, however, always be pollutants that, because of their nature or chemical composition, cannot be contained in such ways. Included in this group are all soluble pollutants such as nitrates, phosphates, chlorides (i.e. de-icing salt) and metals such as copper and zinc.

Reference has been made in Section 4.3.18 to the ability of vegetation to capture certain pollutants out of solution. Further development of biological systems should be encouraged. However, space limitations may not always allow inclusion of such installations.

The only fully effective method which exists at present to counter the threat of soluble and miscible liquids is total containment. This is not a practical solution for pollution from routine discharges, as it would require the controlled disposal of all highway runoff to licensed locations. Containment has, however, been used to provide a defence against accidental discharges. Current practice is not consistent as agreement on many fundamental issues has yet to be reached between highway authorities and regulatory bodies. An inexpensive type of containment facility comprises a short length, typically 20 metres, of open fully lined ditch with a form of baffle at the discharge end to permit flows to be stopped by the simple process of damming the outlet with sandbags. Other systems use gate valves, penstocks, or flap valves, which are suspended open and can be easily released into the closed position. It is important that these can be operated with sufficient speed to nullify the risk of pollution. Such mechanical devices are susceptible to vandalism or mechanical failure at crucial times. A useful rule for highway drainage designers to follow when contemplating a containment provision is to keep it simple and consider the outcome of unauthorised tampering.

4.5 POLLUTION MANAGEMENT

4.5.1 Pollution control through cleaning operations

Street sweeping

Street sweeping in Britain is widely practised and is carried out by local authorities under the provisions of the Environmental Protection Act 1990. This can range from "litter-picks" once or twice a day in shopping areas to mechanical channel sweeps once a year or less in rural locations. The current objectives of cleaning are mainly aesthetic. Sweeping is not currently carried out as a pollution control measure. It is also carried out as a road safety measure, and DOT lays down sweeping frequencies for motorways.

Maintaining and cleaning urban areas can have an impact on the quantity of pollutants washed off by stormwater. The extent of that impact has been the subject of investigation over the last two decades. In a baseline study in the USA, Sartor and Boyd (1972) concluded that the removal efficiency of dirt and debris (sediment) from the street by road sweeping was dependent on the particle size range of the surface sediment, with the small sizes being the least well removed. They pointed out that although the roads were swept three times a week in their study area, the discharge-weighted pollutant concentrations in storm runoff were only reduced by 33% or less.

An experimental catchment was established in Bellevue, Washington, to examine the sources, quantities and removal of urban pollutants. One area was swept three times per week for a five month period and compared with a control area, which was not swept. It was observed that the mean pollutant concentrations in runoff from the swept areas showed minor reductions compared with the control for seven out of the eight constituents studied and only one pollutant showed a reduction as high as 33%. About half the total debris and even less of the material smaller than 63 μm was removed, even though more than half of the suspended material in the runoff was less than this size. Removal of smaller particles was found to be improved by the use of a modified mechanical sweeper (Pitt, 1985).

In Sweden, Malmquist (1978) reported significant improvements in water quality from areas with regular sweeping practices. Runoff from unswept areas contained on average 2.3 times more suspended solids and heavy metals than that from cleaned areas.

Trials in San Jose, California, showed that street cleaning can remove up to 50% of total solids and heavy metal yields in urban runoff, but only if the cleaning frequency is once or twice per day (Pitt, 1979a). Organics and nutrients could not be controlled even with intensive cleaning.

British research in this field has been limited and modest in scope. Ellis (1979) measured the sediment load on a suburban catchment in north-west London and found that only 4-8% of the solids on the street surface were composed of material below 60 μm in size. He reported further that one pass of a mechanical rotary sweeper was only able to reduce the solids fraction below 60 μm by about 15-20%, although there was a near total removal of sediments above 10mm. Ellis (1986) concludes that the cleaning interval is the dominant influence on effectiveness. Achievement of optimum effectiveness requires a sweeping frequency of at least the average time between storms.

A recent CIRIA study (Butler & Clark, 1993) tended to concur with the distribution of surface sediment sizes discussed by Ellis but found cleaning using a mechanical vacuum sweeper to be more efficient for small particle sizes (see Table 4.10). It was considered, however, that removal efficiencies under everyday conditions would be significantly lower, particularly due to obstruction by parked cars.

Table 4.10 Efficiency of sweeping   (Adapted from Butler & Clark, 1993)

Particle size range (μm)	Efficiency (%) Vacuum sweeper	Efficiency (%) Manual sweeper
> 5600	90	57
5600 - 1000	91	
1000 - 300	84	46
300 - 63	77	45
< 63	76	25
Overall average	**84**	**48**

In the same study, a simple lumped-parameter cost model was developed and calibrated, which allowed the input of cleaning cost data, as well as data relating to: sediment build-up rates on the street surface, the frequency and efficiency of the cleaning activities, the efficiency of the gully pot and the effect of rainfall in moving sediments into the gullies and subsequently into the drainage system (Butler et al, 1993). The model showed that increased sweeping frequencies reduced the sediment influx into the system but sweeping was not a cost-effective form of control of suspended solids.

Street flushing

Street flushing (or washing) is not common in Britain but is used in some urban areas for aesthetic purposes, particularly during dry weather (Butler & Clark, 1993). Flushing does not remove particles from the catchment but merely relocates them. The effect of flushing on pollution reduction is marginal to negligible in areas served by separate sewers, since most of the flush is collected by surface water sewers and conveyed to the receiving body of water (Novotny & Chesters, 1981). In many cases the volume of water used for flushing is insufficient to transport the sediment to the nearest gully (Pitt, 1979b). Pravoshinsky (1975) reported that the washings from street flushing is more polluted than ordinary storm runoff.

Street flushing may be advantageous, however, in areas served by combined sewers, as the runoff generated will be conveyed under dry weather flow to the sewage treatment works. In this way accumulated pollutants on the surface will be reduced and hence cause less impact during storm events.

Gully emptying

It has been shown in Section 4.3.2 that gully pots serve a useful function in protecting both the drainage system and the receiving waters from the larger/heavier suspended solids and their associated pollutants. Unfortunately the gully pot can also have a deleterious effect on runoff quality, particularly due to the washout of anaerobic standing liquor.

To ensure adequate operation of the gully pot, it is common for an emptying regime to be employed based on nominal cleaning frequencies of typically once or twice a year (DOT, 1985; LAA, 1989). However, Butler & Clark (1993), in a study of over 130 cleansing operations on ten different sites discovered that only 79% of the scheduled number of gullies were actually cleaned and the measured removal efficiency of the collected sediment amounted to 72%. The impact of current cleaning practice was also monitored and it was found that during cleaning some 10% of the sediment in the gully pot was in fact washed into the receiving stream. It has been noted that, contrary to official guidance, current practice is to carry out flushing and recharging of gully pots with the polluted liquor removed from previously cleaned gully pots. This practice would negate some of the pollution control benefits of pot cleaning. It should be noted that this is, to a large degree, a function of the design of gully pot emptying vehicles rather than lax procedures on the part of the maintenance authorities.

The impact of cleaning operations on pollution has not apparently been quantified. In the USA, Field (1985) recommends that maintenance of the effectiveness of catchbasins (gully pots) for pollutant removal probably requires cleaning twice a year, depending on conditions. Pitt (1985), reporting on the Bellevue trials, suggests that cleaning at that frequency should reduce the lead and total solids concentrations by 10-25%. COD, nutrients and zinc may be reduced by 5-10%.

Work on the speciation and phase changes associated with heavy metals in gully pots (Morrison et al, 1988) led the authors to recommend a cleaning frequency of between 4-7 days to control metal outflow. A rapid reduction in dissolved oxygen in the base of the gully pot, particularly during the summer, has been noted by Fletcher et al (1978), and Pratt & Adams (1984) have suggested that the sump storage volume needs to be doubled to provide adequate oxygen recharge.

The model developed by Butler et al (1993) can also assess the impact of the gully pot in limiting the sediment influx into the drainage system and could predict the relationship between sweeping and cleaning cycles. The role of the gully pot and the efficiency of its cleaning were shown to be particularly important.

Sewer and drain cleaning

Sediment deposition within sewers and drains has been implicated to a greater or lesser extent in the first flush phenomenon observed in stormwater (and combined sewer) discharges (Hall & Ellis, 1985). To ameliorate this problem one option is to control the level of sewer deposits by cleaning. Butler & Clark (1993) reviewed the various methods available and considered their usefulness in sediment management. Typically, it was found to be an expensive option and there is little or no experience of the use of cleaning as a pollution control measure in storm sewers.

In combined sewers, trials have been undertaken of sewer flushing during dry weather where short duration waves of water are introduced into the sewer so as to scour the sediment into suspension and hence transport it towards the sewage treatment works. The US study (Hall & Ellis, 1985) concluded that this technique could transport organics, nutrients and heavy metals sufficient distances to make it a viable pollution control method (Pisano et al, 1979).

4.5.2 Pollution control as part of winter maintenance operations

The effects of de-icing agents such as salt and its substitutes (e.g. urea) on the water environment is discussed in Section 2.3.2. Correct control and management of their use is essential.

Considerable attention has been focused on cost-effective use of de-icing agents in recent years and detailed guidance and advice is given in a number of documents, which include: Department of Transport *Code of Practice for Routine Maintenance of Highways* (1985); *Improving Highway Maintenance: A Management Handbook* (Audit Commission, 1988); *Highway Maintenance: A Code of Good Practice - Winter Maintenance Supplement* (Association of Metropolitan Authorities, 1991); *Spreaders for Winter Maintenance* (BS 1622:1989); *Specification for Salt for Spreading on Highways* (BS 3247:1991) and *Trunk Road Maintenance Manual* (DOT, 1992).

The Audit Commission report indicated that up to 60% of de-icing salt consumption was wasted through over spreading and loss from stockpiles, clearly pointing to scope for both financial savings and reduced salt load to surface and groundwaters. A central theme of current advice is that a great deal can be done at minimum expense to ensure that salt application is carefully managed to produce the required de-icing effect with minimum wastage. This can be achieved through appropriate maintenance and calibration of equipment, the training of operatives and use of the Meteorological Office "Open Road" weather forecasting system, which is especially geared to the requirements of local authorities. In addition to this latter service, the majority of Highway Maintenance Authorities utilise one of two commercial ice protection services. The nett result has been an overall reduction of 10-20% in the number of "outings" compared with practice during the late eighties. This has shown a consequent decrease in salt usage over the same period.

De-icing salt can also be a major source of residual solids entering the drainage system during cold periods. There could, therefore, be benefit in the sweeping/collection of the excess salt after the thaw, although this is currently not done as a matter of routine (Butler & Clark, 1993).

The adverse effects of excessive salt application on roadside vegetation and on reinforced concrete structures are widely recognised. Consequently, considerable attention has been given to finding alternatives to sodium chloride. Experience with the use of urea in the Midlands has not been encouraging because of the wider environmental problems associated with its hydrolysis to ammonia. At present the most promising alternative appears to be calcium magnesium acetate, which has almost the same de-icing effect as salt but is virtually non-corrosive and can be used in existing spreading equipment with little or no modification. The major drawback with this material is cost, which is currently some 15-20 times more than salt. In addition, concerns have been raised about its possible contribution to BOD loadings in receiving waters (Dobson, 1991). It would, therefore, appear that conventional de-icing salt will continue to be used for the foreseeable future but the recognition of its adverse effects on the environment will maintain pressure for the highest standards of management in its storage and use.

4.5.3 Pollution control during other maintenance operations

As noted in Section 2.3.2, the use of herbicides for the control of vegetation on roadside verges and other areas can add to the pollutants in runoff water, particularly if they are used on impermeable surfaces. Consequently, it is important that any application required by maintenance programmes should be carefully controlled and follow appropriate guidance. Specific guidance on the use of sprays is given in the Department of Transport's *Trunk Road Maintenance Manual*, vol. 2 (DOT, 1992), which advocates herbicide use only in exceptional circumstances, or where a particular need exists, such as around posts carrying signs and on kerb edges. All weed spraying should be carried out in accordance with the Control of Pesticides Regulations, 1986, and the Control of Substances Hazardous to Health Regulations (COSHH, 1988), after consultation with the pollution control authority. More detailed advice on specific herbicides can be obtained from ADAS (the executive agency of the Ministry of Agriculture Fisheries and Food that deals with the registration of pesticides).

4.5.4 Pollution control at source

Litter control

Litter and vegetation deposits constitute the major component of street refuse. Litter control, especially in highly impervious areas, can be an effective measure to combat pollution of runoff (Novotny, 1984). Measures can include active public education, provision of more litter bins, more frequent emptying of litter bins and the stricter enforcement of anti-litter laws (Amy et al, 1974). The Environmental Protection Act 1990 has clarified responsibilities with respect to the sweeping of highways. Local authorities have a duty to keep highways clear of litter and standards of cleanliness have been defined (photographically).

Leaf and grass cuttings

Measured and simulated data indicate that during periods of leaf fall, the organic and nutrient pollution from residential areas with trees increases dramatically (Novotny, 1984). Typically leaf fall can amount to 20 kg/tree/year, 90% of which is organic (Novotny and Chesters, 1981). An effective programme for the removal of leaf/grass cuttings from impervious areas should be implemented where possible. Butler and Clark (1993) found that autumn leaf removal programmes were quite common practice among local authorities.

Control of pervious areas

Soil loss from unprotected bare areas can be considerable - up to 100 tonnes/ha/year (Novotny, 1984). Temporary or permanent grass seeding, turfing or mulching should be used on any unprotected surface within a watershed.

Control of construction sites

Construction sites can produce large amounts of pollutants due to stripping of topsoil and exposure of unprotected bare soils. In particular, suspended solids concentrations can be significantly increased. In one Canadian study (Waller, 1986), suspended solids yields were monitored before and during the development of a 14 hectare site and found to increase from 25kg/ha/a to 2100kg/ha/a.

The Control of Pollution Act 1974 contains provisions specifically intended to control construction sites. Techniques that may be used to achieve this control could include temporary protection of exposed soils, minimisation of bare soil exposure, collection and ponding of surface water runoff. Other measures include providing protective skirts around site boundaries, covering-up stockpiles and providing wheel washers for lorries.

5 Guidance

5.1 INTRODUCTION

Sections 2 and 4 of this report give details of the pollution contained in highway runoff and the techniques that are available to highway engineers for the disposal of runoff. This section combines the information from those sections with guidance on how to achieve effective pollution control when designing a drainage system.

The guidelines do not provide a procedure that can give a design for a highway drainage system to meet a defined water quality standard. Until procedures of this type are available, highway engineers are urged to consider the flow quality as being of equal importance as the flow quantity.

The guidelines presented in this section should lead to an improvement in water quality without unjustifiable increase in cost.

5.2 CHARACTERISTICS OF HIGHWAY RUNOFF

5.2.1 Sources of pollutants

The pollution in highway runoff comes from four main sources:

- the operation and passage of motor vehicles; including those arising from abrasion, corrosion and attrition of both vehicles and highway surfaces
- maintenance operations caried out on roads (e.g. de-icing, defoliation)
- accidents and spills
- other miscellaneous sources, e.g. atmospheric deposition, maintenance of vehicles, litter and illegal disposal, agricultural activities.

There are significant differences between the sources that contribute to routine discharges and accidental discharges. Accidental spillages can consist of almost any polluting substance, and their impact will vary from incident to incident. They receive the greatest amount of publicity, but pollution from routine discharges is also important as it occurs almost continuously.

5.2.2 Pollutants of major concern

More than 30 potentially polluting substances have been identified in highway runoff. The effects of these may range from aesthetic nuisance, to causing biochemical deterioration of the receiving body of water. A full discussion of the individual pollutants in highway runoff, with typical concentrations and loadings that can be expected for different highway types, appears in Section 2.

There are two ways in which the impact of pollutants on a receiving water should be assessed. For some pollutants and some receiving waters it is the total amount of a pollutant discharged over a long period of time that is significant. This is particularly the case with discharges to groundwater and with polluted sediments discharged to static surface waters. Here the assessment of pollution can be based on the average annual runoff. For other pollutants discharged to surface waters it is more important to consider the conditions in a single unusual event. For pollution by metals this could be the average over a day, for pollution by ammonia or oxygen demanding substances it could be over one hour.

The pollutants that are of most concern in highway drainage are described here.

Sediments

Sediments are the dominant mass of pollutants from highway runoff. Whilst much of the load is either inert or non-polluting in nature, both the physical loading of sediments blanketing the bed of any receiving body of water and the associated potential for the prolonged release of toxic contaminants can cause chronic, i.e. long-term, pollution downstream. Discharge of sediment may also incur additional costs for dredging to keep the waterway clear. Thus it is desirable that the sediment load should be reduced as much as is reasonably practicable.

The finest fraction of sediment load (<63 μm), which has been reported as only 6% of the total mass, can constitute up to 50% of the pollution load of metals, hydrocarbons, oxygen demand, nutrients and herbicides (Collins and Ridgeway, 1980).

Hydrocarbons

The main hydrocarbons of concern are the petrochemical derived group, which includes petrol, fuel oils, lubricating oils and hydraulic fluids. In unmodified form these are liquid, virtually insoluble and lighter than water. Some hydrocarbons, such as bitumen and heavy fuel oil, become heavier than water when affected by naturally occurring bacteria and can then be treated as sediments. Some hydrocarbons exhibit an affinity for sediments and thus become entrapped in deposits from which they are only released by vigorous erosion or turbulence.

Even low levels of contamination with oil give rise to surface sheens on receiving waters and will also exert an oxygen demand on the receiving water. Hydrocarbons also impart tastes and odours to both surface and groundwaters, which can make them unsuitable for abstraction for drinking water.

Metals

Ten metals have been identified as having a significant presence in highway runoff but only five of these pose a significant threat to receiving waters: cadmium, lead, copper, zinc and iron. The first two are known to be particularly toxic and so might be expected to pose the more serious hazard.

Cadmium is the only one of these five that is on the EC blacklist of dangerous substances. This means that its use for all purposes is controlled and is reducing substantially. It is probable that the current receiving water standards of 5 μg/l is now rarely exceeded by highway discharges so that cadmium is unlikely to be problem in highway runoff now or in the future.

Lead has very limited solubility so that much of the lead in highway runoff can be removed with the sediment load. Very little of the remnant will be soluble and thus its bio-availability and potential eco-toxicity to any receiving body of water will be very low. There has been a substantial reduction in the use of lead as a fuel additive throughout the 80s and continuing into the 90s, so the principal source of lead in highway runoff is rapidly declining toward insignificance. Thus, although lead levels of several mg/l have previously been recorded in highway runoff, all of the studies were conducted during the 70s and early 80s before unleaded petrol had made an impact on the market and so lead in runoff is unlikely to be a significant issue in the future.

It is currently the more commonplace copper and zinc that pose the greater threats. Because they are such ubiquitous metals in the automotive industry and are moderately soluble in water, they appear in highway runoff at significant (mg/l) levels. Copper is an algicide and fungicide at the mg/l level and can be toxic to plants at higher concentrations. Both copper and zinc are toxic to fish at sub-mg/l levels, particularly to fish fry in upland streams, which are of low pH, low alkalinity, low flow and poorly buffered. Both copper and zinc are more soluble and more toxic in low pH waters.

Iron occurs as rust particles, which may have a beneficial effect because the micro-particulate structure and high electrical charge of rust particles cause them to be very effective scavengers and adsorbers of many of the soluble heavy metal ions. However, iron is also present in significant quantities in road salt, and in this form can be a significant pollutant.

Salts and nutrients

Although sodium chloride can be present in runoff at very high concentrations after winter maintenance operations, it does not usually cause an environmental problem because dilution in the receiving water reduces this to acceptable values. The amount of salt that is used is being reduced by better management of winter maintenance, and so the problem is unlikely to become more serious. It may, however, cause a localised problem near to water abstraction points. Salt is very soluble and so is not readily removed by treatment.

Nutrients such as nitrogen and phosphorus may be present at mg/l concentrations and may cause problems of algal blooms or of high nitrate levels preventing water abstraction. However, the bulk of these nutrients is claimed to be adsorbed onto the suspended solids.

Ammonia can be generated in the warm, anoxic environment of gully pot liquors and may then be emitted as a first flush from a gully pot or sediment trap.

The use of urea as a de-icing agent can cause problems of both ammonia and nutrient enrichment with nitrogen.

Others

The main pollutants of concern here are herbicides and pesticides, which can be environmentally harmful at sub μg/l levels. The once ubiquitously employed DDT and organo-chlorine type chemicals are now banned. Atrazine and simazine are currently being phased out and replaced with other substances.

Trace residues of all these chemicals are widespread within the environment. However, only those used currently and in the future need consideration here.

For water to be abstracted for drinking, a combined limit of 1 μg/l is currently in force for atrazine and simazine (DoE, 1989a). The small amount of data available suggest that typical values found in highway runoff are in the sub μg/l level. However, it is possible that a storm immediately following a roadside application could result in peak concentrations above this level. Gomme et al, (1991) found that chemicals, which had initially been held in the soil zone, can subsequently be flushed out by excess water. Removal of the sediment phase might not therefore be the full solution here. It is suggested that further studies of the passage, life and fate of all trace chemicals within a highway drainage system are required.

Summary

From the previous section the pollutants of most concern in highway runoff are:

- sediments
- metals (zinc, copper and iron)
- hydrocarbons (oil and petrol)
- pesticides and herbicides.

Hydrocarbons are a problem at very small concentrations, and it is difficult to predict their concentration in the discharges. These are best dealt with by the use of appropriate technology, such as oil separators, in areas where they may cause a problem.

Pesticides and herbicides can cause problems at only trace levels, and the best method of dealing with this is by changing practice so as to reduce their use and to use only degradable chemicals, which do not cause a long-term problem.

5.2.3 Quantities of pollutants

Section 2 presented all of the available data on the build-up rates for pollutants from the normal passage of traffic. From this data the following table has been derived giving typical values for the important pollutants for various traffic densities. The quoted values are if anything an overestimate as they are biased towards the higher values quoted in the previous studies.

Table 5.1 Typical pollutant build-up rates (kg/ha/a)

Pollutant	Total solids	COD (kg O_2)	NH_4-N	Copper		Zinc	
Traffic flow AADT				Total	Soluble	Total	Soluble
< 5000	2500	250	4.0	0.4	0.2	0.4	0.2
5000 - 15 000	5000	400	4.0	0.7	0.3	1.0	0.5
15 000 - 30 000	7000	550	4.0	1.0	0.4	2.0	1.0
> 30 000	10 000	700	4.0	3.0	1.2	5.0	2.5

Section 2 also presented information on spreading rates for salt and the polluting content of salt. This is summarised in the following table.

Table 5.2 Pollutants per outing of road salting (kg/ha)

	Sodium Na^+	Chloride Cl^-	Iron Fe
Precautionary salting	35	55	0.15
Snow or ice	140	220	0.60

5.2.4 Washoff of pollutants

The worst case for washoff of routine pollutants is the short duration, intense summer storm. Intense rainfall is the most effective for mobilising pollutants as the impact of the raindrops removes accumulated pollutants from road surfaces, and moves contaminated sediments from inside gully pots and drainage systems. A short duration storm has a small total rainfall volume, thus reducing the available dilution with consequent increases in concentrations. Summer storms usually occur with low water levels in the receiving water so that dilution is reduced leading to higher residual concentrations. Summer storms also often have significant dry weather periods beforehand during which pollutants can build up on the road surface and in the drainage system.

The first runoff generated by a storm tends to carry a high contaminant load (the first flush). This is due to pollutants that have accumulated on a road surface and in the drainage system during an antecedent dry period being washed off by the initial runoff. This has been associated especially with gully pots, which may have generated anoxic conditions during the antecedent dry period. Thus the first volumes of water displaced from them can be high in soluble organic compounds, BOD, COD, NH_4 and bacteria. In some situations this short-term high concentration will require special consideration.

The worst case for washoff of the residue of winter salting is a thaw after a continuous period of cold weather with several successive applications of salt.

5.3 WATER QUALITY STANDARDS

5.3.1 Framework

The legal basis for the setting of water quality objectives and water quality standards is described in Section 3. This section defines the standards, which are currently in use, that are relevant to highway drainage discharges.

The basic philosophy is that each receiving water has a water quality objective, that is a required use for which the water must be suitable. Typical uses include: drinking water abstraction, salmonid fishery etc. There may be more than one objective for a receiving water and the water must be suitable for all of these uses. It is proposed that these objectives will eventually be set as statutory requirements which must be met by the pollution control authorities. At present they are set as policy objectives by these authorities. For England and Wales the policy is to maintain the possible uses that are currently met by each receiving water.

Each defined use will have a set of water quality standards, such as maximum pollutant concentrations, which must be met for the water to be suitable for that use. These standards apply to the receiving water rather than to discharges into it. The standards may be set as average annual concentrations, concentrations that are exceeded for only 5% or 10% of the time, maximum allowable instantaneous concentrations, or as tables of concentrations for various durations of exceedance and frequencies of occurrence.

The pollution control authorities can ensure that the receiving water meets its target standards by regulating the discharges that are made into it. For most discharges this is by a system of consents, which can specify the quantities and qualities of discharges that can be made. Highway drainage discharges are normally exempt from consents, unless it can be shown that they are causing pollution. In this case a prohibition or consent can be applied. The aim of consents is that the receiving water will meet its standards, but the way in which the consent is set and the standards set in it are not necessarily the same as the standards required in the receiving water. For example, if a large dilution is available, then the concentrations permitted in the discharge may be higher than the river quality standards. Alternatively, the consent may be set as simply requiring the use of an agreed pollution abatement technology, for example an oil separator, rather than limiting the quantity or quality of the discharge. In complex situations it may be necessary to consider all other discharges into the receiving water, and the self-purification capacity of the water in order to determine the conditions that should be set on a consent to discharge.

5.3.2 Groundwater

The policy for water quality objectives and standards for England and Wales for groundwaters is set out in *Policy and practice for the protection of groundwater* (NRA, 1992).

The factors that combine to influence the vulnerability of groundwater to pollution can be summarised as:
- the nature of the water bearing strata
- the presence and nature of drift deposits overlying the strata
- the presence and nature of the overlying soil
- the depth of the unsaturated zone.

The policy distinguishes between protecting groundwater resources, whether they are currently used or not, and of protecting individual major sources where groundwater is abstracted for use.

Resource protection

The policy is that all groundwater should be protected, but the degree of protection depends on the possible use of the water. Groundwater resources are classified as: non-aquifers, minor aquifers and major aquifers.

The policy states that discharges to soakaways from major roads will require oil separators for all three classes.

In addition, for both major and minor aquifers, the policy requires an investigation to be made of the risk of contamination and additional conditions set as appropriate. Long-term monitoring may be necessary to ensure that contamination does not occur.

Source protection

Source protection requires the definition of source protection zones around each major abstraction point. Three zones are defined:

1. Inner zone. This is the part of the aquifer from which water can travel from any point below the water table to the source within 50 days. The minimum size for this zone is 50 m around the source. If the aquifer is confined beneath a substantial cover of low permeability then this zone is not defined.

2. Outer zone. This is the part of the aquifer from which water can travel from any point below the water table to the source within 400 days. As for zone 1 it would not be defined for a completely confined aquifer.

3. Catchment zone. This is the area needed to support abstraction at the source from long-term average annual recharge. All groundwater within this zone will eventually travel to the source. For confined aquifers the catchment may be some distance from the source.

In aquifers that are dominated by fissure flow (such as chalk and limestone) and where high flow speeds are possible (Price et al, 1989, 1992), zones 1 and 2 may cover the entire catchment zone.

The NRA is drawing up source protection zone maps, but these are sometimes based on limited information, and the situation may change as patterns of abstraction change. In some circumstances it may therefore be necessary to make a detailed assessment of the extent of the zones in the area of a proposed highway scheme.

The NRA policy for highway drainage in the source protection zones is summarised below.

For zone 1, discharges of drainage from major roads will not be accepted to soakaways, and the construction of new major roads, however they are to be drained, will be opposed.

For zone 2, discharges of drainage from major roads will be permitted to soakaways only in exceptional circumstances and if detailed investigation shows that there will be minimal pollution.

For zone 3, the conditions are similar to those for resource protection.

For a complete statement of the policy the NRA document should be consulted. The local pollution control authority should always be consulted for advice on how the policy applies in each individual case.

Standards

The policy of protecting groundwater resources requires that discharges into them should not degrade the existing quality of the water. Therefore, for resource protection and zone 3 protection, the average annual concentration of pollutants in the discharge should be no greater than that existing in the aquifer. Where this is unnecessarily onerous because the aquifer is of very high quality, then the discharges should meet the standards for potable water to avoid long-term degradation of the aquifer below this standard. An abbreviated form of these standards is given in Table 5.3, which includes only the pollutants relevant for assessing highway drainage (DoE, 1989b). However, it should be noted that in many cases the discharge will only be required to match the existing quality of the aquifer and that these concentrations may be significantly higher than those in the table.

It can be assumed that in most situations only the dissolved phase of the pollutants will enter the groundwater, with any pollutants attached to sediment remaining in the drainage system, or in the ground immediately around the soakaways. Only the dissolved pollutants therefore need to be considered for comparison with the standards, provided that there is no danger of the pollutants attached to the sediment being subsequently dissolved.

Table 5.3 Standards for water for human consumption

Parameter	Units	Concentration
Nitrate	mg/l NO_3	50
Dissolved iron	μg/l Fe	200
Copper	μg/l Cu	3000
Zinc	μg/l Zn	5000
Cyanide	μg/l CN	50
Lead	μg/l Pb	50
Chloride	mg/l Cl	400
Phosphate	μg/l P	2200

5.3.3 Surface waters

Conditions in surface waters change more quickly than in groundwaters, and they are generally sensitive to short-term peak concentrations. Most standards are therefore quoted as 90 percentile (90%) or 95 percentile (95%) values. The 90% value is the concentration that is only exceeded in 10% of samples which are taken at regular intervals. It can therefore be considered to be the concentration that is exceeded for only 10% of the time. Similarly, the 95% value is the concentration that is only exceeded for 5% of the time.

Use for water abstraction

Where water is to be abstracted for a supply of drinking water, standards are set for this use. Three levels of standard are set depending on the level of treatment that is provided to the water (DoE, 1989a).

DW1   Physical treatment and disinfection
DW2   Normal physical and chemical treatment and disinfection
DW3   Advanced physical and chemical treatment and disinfection

The table of standards that is given below is restricted to the pollutants that are present in significant quantities in highway drainage. All values are 95% values.

Table 5.4 Standards for abstraction of drinking water

Parameter	Units	DW1	DW2	DW3
Nitrate	mg/l NO_3	50	50	50
Dissolved iron	µg/l Fe	300	2000	-
Copper	µg/l Cu	50	-	-
Zinc	µg/l Zn	3000	5000	5000
Lead	µg/l Pb	50	50	50
Cyanide	µg/l CN	50	50	50
Total ammonia	mg/l N	0	1.5	4

Fisheries use

The other main use that is recognised for surface waters is to maintain a ecosystem for fisheries. Rivers have been classified, on a regular basis, using the National Water Council (NWC) scheme. This had four main classes from 1 *High quality* to 4 *Bad quality*. The two highest classes were subdivided giving six classes in total. Each river has a current class and some rivers also have a river quality objective of a class higher than their current one.

The new framework of Water Quality Objectives introduces six classes of fisheries ecosystem (FE) (DoE, 1993). These are at present equated to the NWC river classes and are:

FE	NWC river class	Description
1	1a	High class salmonid and cyprinid fisheries
2	1b	Sustainable salmon and high class cyprinid fisheries
3	2a	High class cyprinid fisheries
4	2b	Sustainable cyprinid fishery
5	3	No sustainable fishery
6	4	Fish unlikely to be present

Table 5.5 Standards for fisheries ecosystems

Parameter	Units	Type	Hardness mg/l $CaCO_3$	1	2	3	4	5	6
Dissolved oxygen	% sat	10%		80	70	60	50	20	-
BOD (ATU)	mg/l	90%		2.5	4.0	6.0	8.0	15.0	-
Total ammonia	mg/l N	90%		0.25	0.6	1.3	2.5	9.0	-
Un-ionised ammonia	mg/l N	95%		0.021	0.021	0.021	-	-	-
pH minimum		5%		6	6	6	6	-	-
pH maximum		95%		9	9	9	9	-	-
Dissolved copper	µg/l Cu	95%	<= 10	5	5	5	5	-	-
			> 10 & <= 50	22	22	22	22		
			> 50 & <= 100	40	40	40	40		
			> 100	112	112	112	112		
Total zinc	µg/l Zn	95%	<= 10	30	30	300	300	-	-
			> 10 & <= 50	200	200	700	700		
			> 50 & <= 100	300	300	1000	1000		
			> 100	500	500	2000	2000		

Additional standards for fisheries ecosystems have been set as guidelines for short-term impacts on dissolved oxygen and ammonia. These are intended for assessing waters receiving discharges from combined sewer overflows or sewage treatment works. For waters that receive these discharges in addition to road drainage, or where there are particular problems of ammonia or reduced dissolved oxygen levels, then these standards should be used in addition to those given above. The standards are expressed as tables of concentrations that should not be exceeded for specified durations and frequencies. The standards are published in the UPM Manual (1994).

5.4 GUIDE TO GOOD PRACTICE

The polluting impact of a road increases with its traffic density. This is both a direct increase for the pollutants deposited by traffic, and an indirect increase for pollutants from road salt where the road width and frequency of salting are likely to increase with traffic flows.

Minor traffic routes do not generate sufficient pollutant load to require any specific treatment.

For medium sized roads there may be some impact on receiving water quality. Much of the problem can be avoided by good choice of drainage system and by removing hydrocarbons from the discharge using oil separators.

For major roads or other roads in particularly sensitive areas then some form of treatment may be required.

The method of collection and conveyance of the drainage should not be a prejudged decision. Unless there are extenuating circumstances, the design of the system should be considered after the pollution control requirement has been assessed. In all cases the need for special precautions to deal with accidental spillages should be considered. These can only be tackled effectively by use of some form of containment facility. The highway engineer should satisfy himself that if no provision is made it is because the risks have been considered fully and are acceptable to all authorities concerned.

5.4.1 Groundwaters

The pollution control authorities may require the selection of a suitable system to treat the runoff prior to discharge to soakaways, infiltration trenches or basins. They should always be consulted at the earliest opportunity. The requirements of the authority depend on the zone into which the discharge point falls. Figure 5.1 shows the procedure to be followed to assess these discharges. The following notes give details of the requirements of the pollution control authorities for each of the zones.

Source protection zone 1

Infiltration of highway drainage will not be acceptable.

Source protection zone 2

There is a presumption against infiltration and it will be acceptable only in exceptional circumstances after detailed hydrogeological investigation. The circumstances in which it might be permitted is if the infiltration would be into a thick unsaturated zone of low permeability that will provide protection to the aquifer. However, these conditions are unlikely to be suitable for infiltration drainage. Oil separators, containment devices and other precautions will be required.

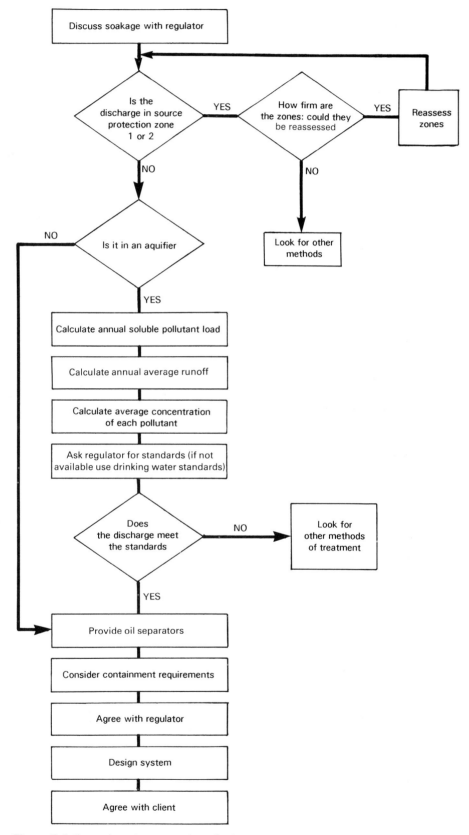

Figure 5.1 *Procedure for assessing discharge to groundwaters*

Resource or catchment zone protection

Infiltration of highway drainage may be permitted following investigation. Oil interceptors will be required, and containment for spills may also be necessary. A suitable form of initial investigation is as follows:

1. Consult the pollution control authority for water quality standards that must be met in the receiving groundwater. If no other information is available, then the standards for drinking water given in Table 5.3 should be used.

2. Calculate the annual soluble pollutant load of copper and zinc in kg for the traffic flow and road area using Table 5.1. Estimate the annual load of chloride and iron from winter salting using Table 5.2.

3. Estimate the average annual runoff from the highway catchment (m^3) by multiplying the average annual rainfall (Figure 5.2) by an appropriate runoff coefficient and the area of the road. If no better information is available then a runoff coefficient of 0.5 should be used.

4. Divide 2. by 3. to obtain the average concentration of each pollutant in the soakage. If this is less than the required standards then the normal runoff poses no threat to the aquifer.

5. If it is higher then consider forms of treatment or alternative disposal methods.

5.4.2 Surface waters

The procedure to follow for assessing the requirements for discharges to surface waters is set out as a flowchart in Figure 5.3 and is explained in the notes which follow. The basic procedure considers the impact of dissolved metals, specifically copper and zinc, and of oil. In special situations, such as drinking water abstractions or static waters, other steps are included to assess other pollutants.

The assessment of dissolved metals is based on the numerical method set out in Appendix A and has a sound basis in water quality standards. The assessment of oil is more difficult. Oil is both an aesthetic pollutant, causing unsightly surface sheens and also has harmful effects on the river ecosystem. There are no water quality standards for oil which take into account the aesthetic effects. The guidance given here for the provision of oil separators is, therefore, based on experience of the types of situation in which they have been found to be desirable. This guidance may not be applicable in all situations and the local impact of aesthetic pollution at each site should be considered.

The impact of dissolved metals and oil depends on the amount by which they are diluted by the receiving water. Each receiving water can be considered to have a dilution capacity to assimilate a certain amount of discharge. The dilution cannot be assessed solely for one outfall in isolation, because some of the dilution capacity may be used up by other adjacent discharges from the same road, or other roads. The flowchart therefore goes through two steps - one to assess the impact of an individual outfall, and the second to assess the effect of all adjacent outfalls.

A worksheet to assist with carrying out the calculations involved in the procedure and a worked example are given in Appendix B.

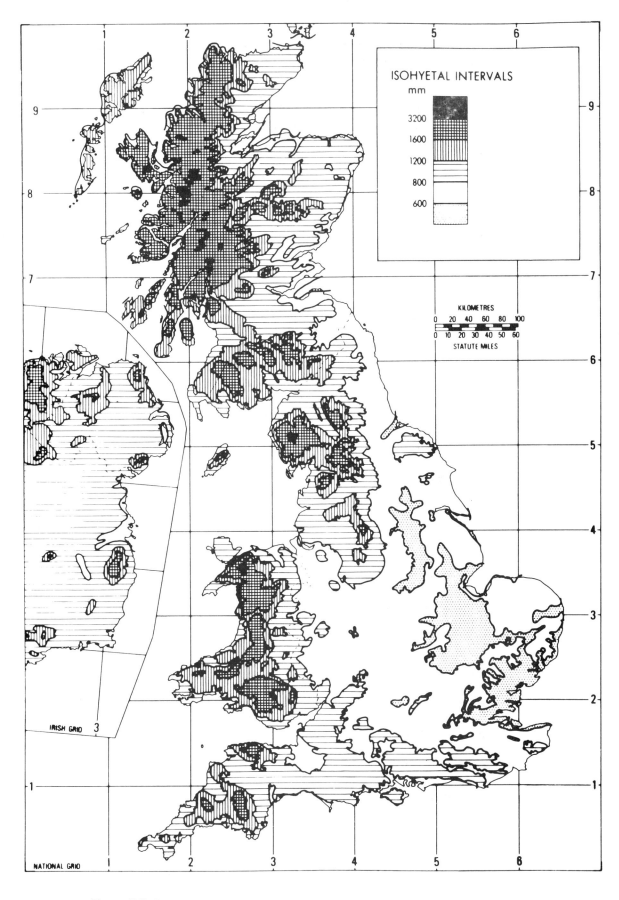

Figure 5.2 *Average annual rainfall 1941-1970*

CIRIA Report 142

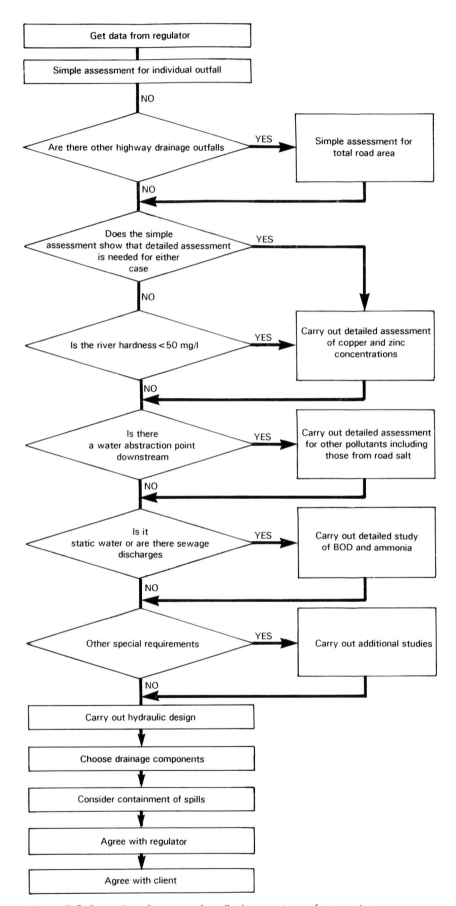

Figure 5.3 *Procedure for assessing discharges to surface waters*

Get data from regulator

Ask the pollution regulator for the following information on the river:

1. What length of the river should be taken as a reach for assessing the combined effect of all highway drainage outfalls?
2. 5 percentile low river flow (Q_5) in m^3/s.
3. River class.
4. Average hardness in mg/l $CaCO_3$.
5. Are there any drinking water abstractions downstream of the discharge point?
6. Are there sewage discharges such that BOD and ammonia are critical pollutants?
7. Are there any other special requirements, such as an SSSI downstream of the discharge point or a discharge to a static water?
8. Additional information on background concentrations may be required later.

If the discharge is to a canal, British Waterways should also be consulted and may have additional requirements.

Simple assessment for individual outfall

1. Read Figure 5.4 for the appropriate depth of rainfall R in mm.
2. Calculate the runoff volume (V) from the highway (m^3) by multiplying by an appropriate runoff coefficient and the area of the road. If no better information is available then a runoff coefficient of 0.5 should be used.
3. Calculate the daily volume of river flow (m^3) at low flow conditions.
4. Calculate the dilution as runoff volume divided by volume of river flow.
5. For this dilution and the design traffic flow of the road consult Table 5.6 for the pollution abatement measures that will be required.

Simple assessment for all contributing roads

It is also necessary to check the combined effect of all adjacent highway drainage outfalls to allow for the river dilution that is already taken up by these other discharges.

1. Identify the length of the river that should be considered (the reach). This will be the length between major tributaries or other sources of inflow.
2. Estimate the total area of roads with traffic flows greater than 5000 veh/day that drains to this reach.
3. Calculate the dilution for the runoff from this road area as described above.
4. Estimate the average traffic flow on these roads.
5. Consult Table 5.6 to see if additional pollution abatement measures are required on the new scheme.

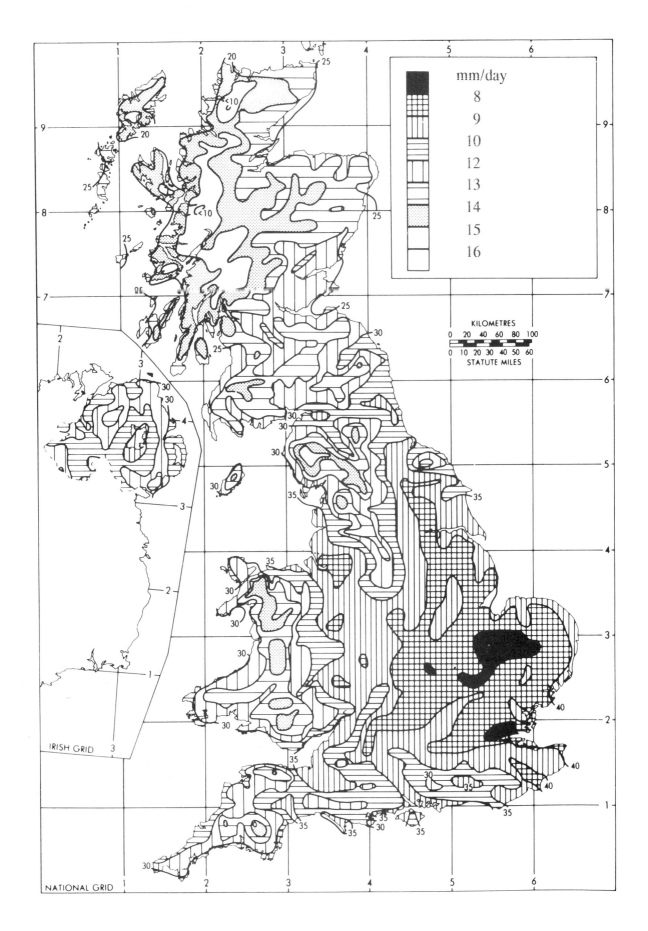

Figure 5.4 *Depth of rain for assessing pollutant washoff*

Table 5.6a Dissolved pollutant abatement requirements

Traffic count veh/day	Fishery ecosystem 1 - 4					
	Dilution					
	2	3	4	6	12	16
< 5000						
5000 - 15 000	D					
15 000 - 30 000	D	D				
> 30 000	D	D	D	D	D	

KEY   D   Assessment of copper and zinc dilution required

Table 5.6b Aesthetic pollution abatement requirements

Traffic count veh/day	All fishery ecosystem classes						
	Dilution						
	2	3	4	6	12	16	18
< 5000							
5000 - 15 000	O	O					
15 000 - 30 000	O	O	O	O			
> 30 000	O	O	O	O	O	O	

KEY   O   Oil separation may be desirable

Detailed assessment of copper and zinc concentrations

1. Identify all roads with a traffic flow of more than 5000 vehicles per day that drain to the river reach.

2. Estimate the area of these roads for each of the traffic flow groups.

3. Calculate the total runoff volume and dilution from these roads as described above.

4. Calculate the buildup over five days of total zinc and soluble copper on each of the road classes using the figures in Table 5.1. Use this to calculate the buildup in five days for all of these roads (M in kg).

5. Obtain the background concentrations (C_b in mg/l) of total zinc and soluble copper upstream of the discharges from the regulator. If it is impossible to obtain these, assume that they are half of the values given in Table 5.5 for the appropriate river class and hardness. The accuracy of the method will be reduced if this assumption is made.

6. Resulting concentration in the river in mg/l is:

$$C_r = C_b + \frac{1000M}{V(1 + Dilution)}$$

7. If these calculated figures are greater than the standards in Table 5.5 then removal of fine sediment and dissolved pollutants are required. See Section 5.5 for details of the drainage system components that can do this. The percentage reduction of pollutants that these will provide should be estimated and the calculation above repeated, reducing the pollutant mass M by this percentage, to ensure that the standards will be met.

Detailed assessment of other pollutants when there is abstraction for drinking water downstream

The additional pollutants that need to be assessed are primarily chloride and iron from road salt. This needs a different method of calculation from that described above. The calculation would be carried out for the combined effect of all outfalls draining into the river reach.

1. Estimate the maximum number of consecutive applications of road salt in one period of cold weather in a typical year. Local information will be required for this.

2. Using the information in Table 5.2, calculate the total load of chloride and iron that would be deposited in this period (M in kg).

3. Calculate the runoff volume and dilution as described above, except use average flow rather than 5% low flow.

4. Obtain background average concentrations (C_b in mg/l) for the river upstream of the discharges from the regulator.

5. Resulting concentration in the river in mg/l is:

$$C_r = C_b + \frac{1000M}{V(1 + Dilution)}$$

6. If these calculated figures are greater than the standards for chloride in Table 5.3 or for iron in Table 5.4 then removal of fine sediment and dissolved pollutants are required. See Section 5.5 for details of the drainage system components that can do this. The percentage reduction of pollutants that these will achieve should be estimated and the calculation above repeated, reducing the pollutant mass M by this percentage, to ensure that the standard will be met.

Detailed assessment of BOD and ammonia

Where the river is already subject to discharges from sewage treatment works or combined sewer overflows then the additional oxygen demand and ammonia concentration in highway drainage may be sufficient to degrade the river quality. The studies required to assess the effect of these pollutants are complex as the pollutants decay with time and both have an effect on the dissolved oxygen content of the water.

A full description of this type of study is beyond the scope of this report but is described in the Urban Pollution Management manual (UPM, 1994).

Other special studies

Other special studies may be required if there is an SSSI downstream of the discharge, or if the discharge is to static waters where the buildup of sediment or nutrients would cause problems. The regulatory authority may specify which pollutants are to be considered. These would normally be considered either by calculating the total annual discharge or in a similar way to copper and zinc using the procedure defined above.

5.5 APPROPRIATE DRAINAGE TECHNIQUES

Once the requirements for the quality of the discharge have been established the system can be designed and the method of drainage decided on. The components of highway drainage systems fulfil one or more of the following functions:

- *collection* of water from the road surface or sub-surface
- *conveyance* of surface water to another point in the system
- *disposal* of collected surface water to watercourses or groundwater
- *storage* of surface water to reduce peak flows
- *coarse sediment removal* to prevent blockages
- *pollutant removal* to protect receiving waters.

Pollutant removal may be further subdvided into the removal of pollutants attached to fine sediments, the removal of dissolved pollutants and the removal of oil.

It is useful to carry out the hydraulic design of the system at this stage, as this may indicate that attenuation storage is needed in order to meet flow restrictions on the discharge to surface waters. Storage ponds can also be effective in the removal of sediment and associated pollutants and so may avoid the need for other measures.

The following tables show which drainage methods and components are suitable to provide various levels of treatment, and that are appropriate and cost-effective in various locations.

After the tables there are notes on the use of several of the options.

5.5.1 Gully pots

Trapped gullies consisting of gratings and gully pots will continue to fulfil an important role in the collection and pre-treatment of highway runoff, particularly in the urban environment where safety and amenity requirements dictate the use of kerbs.

The removal of coarse sediments is a vital function both in its own right and because solids are carriers of much of the pollutant load. Gully pots can be considered, if properly maintained, to retain effectively solids greater than 300 μm in diameter. Smaller particles may be partially caught but are liable to agitation and suspension during heavy rainfall.

Soluble pollutants are not significantly reduced and indeed may be increased in the liquor during periods of dry, warm weather. Small quantities of oil and other hydrocarbons can be retained in gully pots equipped with traps, under low flows, but are readily washed out if maintenance is inadequate. Gully pots do provide some very limited emergency storage of spillages.

The design of the gully pot as an instrument for pollution control is still in its infancy and there are no established procedures at present. In principle the gully pot should be chosen (from the standard sizes available) for its solids capture efficiency and the sump should provide sufficient sediment storage space given the actual cleaning regime. It should be emphasised that unless gully pots are cleaned regularly and thoroughly their benefit will be much reduced.

5.5.2 Combined surface water and groundwater filter drains

Filter drains have historically been used for draining surface runoff from carriageways, verges, cutting and embankment slopes and adjacent land, together with a sub-grade drainage function. A particular application has been for roads in cuttings. In their guidance document HA39/89, the DOT have made recommendations that their use should be limited to instances where resultant economies can clearly be identified, and they are not at present generally advocated by the DOT.

Management measure	Type	Pollutant removal effectiveness				Relative capital costs/ha.	Additional land requirements	Operational and maintenance costs	
		Fine sediments < 63 μm	Metals	Herbicides	Organics			Routine	Non-routine
Informal verge	Collection	M to H	H	H [1]	H	L	L to M	O	L
Grassed ditches	Conveyance	H	H	M	H	L	L	L	L
Swales	Conveyance/storage	H	H	M	H	M	M	L	L
Gully pots	Collection	L	L	L	L	M to H	L	H	H
Filter (french) drains	Collection/conveyance	M to H	H	H	H	M to H	L	O	H
Infiltration basins	Disposal	H	H	M	H	H	H	H	H
Soakaways and infiltration trenches [2]	Disposal	H	H	M	H	H	H	H	H
Storage ponds (wet detention)	Storage	M	H	H	H	H	H	L	L
Detention tanks (dry detention)	Storage	L to M	L to M	L to M	L to M	M	L	L	L
Catchpits and grit traps	Conveyance	L	L	L	L	M to H	M	M	M
Oil separation	Conveyance	L	L to M	L	L	H	M	H	H
Sedimentation tanks	Conveyance	L to M	M to H	L	L to M	H	M	H	H
Wetlands	Storage/disposal	H	H	M to H	M to H	H	H	O	M
Lagoons	Storage	M	H	M	L to M	M to H	H	M	H
Infiltration pavements	Collection/storage	L to M	H	N/A	H	H	O	M	H
Operational systems									
Litter control	Source	L	L to H	L to H	L to H	L	O	O	O
Control of de-icing salt / chemicals	Source	L	H	H	H	L	O	L	O
Control of herbicides / pesticides	Source	N/A	H	H	H	L	O	L	O
Street sweeping / cleaning	Source	L	L	L	L	M to H	O	H	H

Ratings: H = high, M = medium, L = low, O = none, N/A = not applicable
[1] refer to 5.5.3 for recommendations on vegetation control   [2] refer to 5.3.2 for recommendations on groundwater

Table 5.7 Ratings table for management measures

(based on an idea by Maestri et al, 1989)

Drainage/pollution measure	New highway				Retrofit or upgrade of highway			
	Interchange	On embankment or elevated	At-grade	In cutting	Interchange	On embankment or elevated	At-grade	In cutting
Informal verge [1]	L	H	M	N/A	L	H	M	N/A
Grassed ditches [2]	H	H	H	N/A	H	H	H	N/A
Swales	H	M	M	N/A	M	L	L	N/A
Gully pots	H	H	H	H	H	H	H	H
Filter (french) drains	M	H	H	H	M	H	H	H
Infiltration basins [2]	H	L	H	L	H	L	H	N/A
Soakaways and infiltration trenches [2]	H	L	H	N/A	H	L	H	N/A
Storage ponds (wet detention)	H	L	H	L	H	L	M	L
Detention tanks (dry detention)	H	L	H	L	H	L	M	L
Catchpits and grit traps	H	H	H	H	H	H	H	H
Oil separation	H	H	H	H	H	H	H	H
Sedimentation tanks	M	M	H	M	L	L	M	L
Wetlands	H	L	H	N/A	H	L	L	N/A
Lagoons	H	L	M	L	H	L	L	N/A
Infiltration pavements	L	H	M	L	L	H	M	L

Applicability Ratings: H=high, M=medium, L=low, N/A=not applicable
[1] refer to 5.5.3 for recommendations on vegetation control   [2] refer to 5.3.2 for recommendations on groundwater

Table 5.8 Applicability of drainage measures to highway situations

However, there is a good case for the wider application of filter drains, particularly in view of their good removal performance for sediments and associated pollutants (Table 4.3). In some circumstances there will be no obvious practical alternative, especially when the uncertainties concerning the long-term effectiveness of fin drains (Section 4.3.8) are considered. For reconstruction work and where large groundwater flows from cuttings are to be dealt with, filter drains are still included in DOT relevant design guidance (Advice Note HA39/89, 1991).

A further refinement worthy of consideration for roadside or near-road locations that are primarily intended to receive and remove surface runoff is to line the trench itself with an impermeable membrane. This modification, which was included in the M1 study (Section 4.3.3) for experimental purposes, potentially eliminates or severely reduces egress into the sub-grade, but retains the treatment characteristics of the stone fill whilst permitting conveyance of the collected water to a surface water discharge where appropriate. Issues of effective membrane installation and effective working life need to be considered carefully however.

One solution to some of the problems associated with filter drains is shown in Figure 5.5.

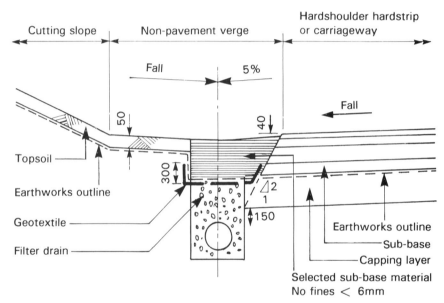

Figure 5.5 *Typical cross section of an improved filter drain*

5.5.3 Informal verge systems

The informal verge system has its main application in rural locations. From a runoff quality enhancement point of view it has features to commend it. It has recognised shortcomings but these are principally related to over-running by vehicles and can be overcome by the use of geotextiles to provide reinforcement.

Provided highway run-off can be effectively drained onto the verge by careful edge of carriageway detailing, the grass cover can provide pollutant-reduction mechanisms, similar to those encountered in swales. This can lead to reductions in both solid based and soluble pollutants.

The effect of the verges can be enhanced in a number of ways, by:

- establishing dense, salt tolerant grass cover, wherever possible
- limiting the number of grass cuts per season, commensurate with safety requirements
- leaving grass cuttings on the ground to act as additional filter material and reduce flow velocity, although this is not recommended in an urban situation.

5.5.4 Infiltration basins

Unlike filter drains, infiltration basins are generally designed to be sited some distance from the road and runoff is conveyed to them. Consequently they should have minimal adverse impacts on nearby pavement foundations, but their potential impact on sensitive groundwaters is probably similar to filter drains and soakaways. Considerable misgivings over their long-term performance and potential maintenance costs, as noted in Section 4.3.13, indicate that they should not be regarded as a first choice runoff treatment system. Further research to establish a better understanding of their performance and efficiency may change their current image.

5.5.5 Swales (grass channels)

As reported in Section 4.3.12, current British design approaches for swales (CIRIA, 1992; Ellis, 1991b) are based principally on optimising the hydraulic efficiency of surface runoff flows (i.e. high flow velocities, minimisation of erosion). However, to achieve water quality improvement (by bio-filtration, sedimentation, etc.) sufficient contact time must be provided and this means basing the swale capacity on an upper limit of flow velocity.

Presently, very little information is available that could be considered as an established design basis for achieving significant water quality improvements. Some guidance is provided by Ellis (1992a) who has recommended that the swale design should be based on the following:

- vegetation height equal to the design flow depth
- maximum flow event of 1:1 return period
- maximum velocity should not exceed 0.8 to 1.5 m/s
- a Manning value of about 0.05.

Further advice on other aspects of design criteria and maintenance practice for improving water quality are given in Ellis (1992a) and Ellis (1991b).

5.5.6 Soakaways and infiltration trenches

The widespread use of soakaways for disposal of many types of runoff water has been noted previously (Section 4.3.14), but, as with filter drains and infiltration basins, their use as a means of disposal of highway runoff to ground is increasingly called into question. Where some form of infiltration is unavoidable, on roads carrying little traffic, and where risks of groundwater contamination are minimal, filtration trenches may offer a more effective means of disposal, because of the larger surface area available for infiltration. For both types of structure there are recommendations for improved designs in BRE Digest 365 (1991) and a detailed manual of infiltration techniques will be published by CIRIA (RP448).

5.5.7 Storage ponds and detention tanks

In general, flow detention in storage ponds and detention tanks will lead to particle settling and removal, together with its associated contaminant load, and some bacterial die-off and soluble contaminant reduction (e.g. metals, nutrients) may occur. The extent of the treatment achieved will depend on the type of storage pond (e.g. dry or wet), the mean flow detention time, and the pond design.

For all types of storage ponds/tanks the minimisation of turbulence and hydraulic short-circuiting, and the increase of flow detention time, will improve overall pollutant removal. For wet ponds, Ellis (1991b) has suggested that the maximum pollutant removal appears to be produced when the pond volume/mean total storm runoff volume ratio is in the range of 4 to 6 (for detention volumes $> 100m^3$ per effective hectare) and when the ratio of the pond surface area to the drained road surface area is in excess of 2 to 3%. A substantial removal of total solids and solid-associated pollutants ($>50\%$) can be expected where the average retention time is from 12 to 16 hours (Ellis, 1992a).

The experience to date with dry storage ponds/tanks suggests they have a low efficiency in pollutant removal due principally to the problem of re-entrainment of solids during high flows. Satisfactory measures to counter this have not yet been established.

Additional methods of increasing pollutant removal that may be worth considering are:

- grit/oil chambers located at the inlet
- forebay or diversion structures at the inlet to minimise the initial flush of pollutants
- marginal planting of emergent macrophyte plants
- provision of deep water zones (>2.5m depth) to encourage algal growth and provision of fish.

These aspects are described in more detail in Ellis (1992a).

5.5.8 Catchpits/grit traps

Catchpits have been used widely on highway drainage schemes. Larger grit traps, however, are less common in British practice. The main application of catchpits is as a replacement for several gully pots, thus reducing the extent of maintenance required. Consequently, the quality implications are similar to those noted in Section 5.5.1. As for gully pots, these pits and traps will only (at best) be effective in retaining coarse sediment.

Design principles for catchpits, and particularly grit traps, should follow good hydraulic practice. That is, ideally, flow should be as quiescent as possible with mean velocities less than 0.3 m/s. It is important too that adequate means of access and solids removal are provided.

5.5.9 Oil separation

Oil removal from highway runoff is more important for heavily-trafficked roads that drain to sensitive surface or groundwaters. Consequently, oil separators are being increasingly recommended for new major road developments although their use is at present limited. Essential design factors to be considered have been referred to previously (Section 4.3.11) but it is also important that the need for regular maintenance and associated vehicular access is taken into account at the design stage.

Oil separators are suitable for both rural and urban locations.

5.5.10 Sedimentation tanks

In the comparative M1 study of a filter drain, sedimentation tank and lagoon, reported in Section 4, the 1500 l sedimentation tank was found the least effective treatment system of the three, although not surprisingly it occupied a much smaller land area than the other two systems. In principle, therefore, such an arrangement could be considered when a modest level of solids removal is required but where space is very limited. In most circumstances, however, similar solids removal efficiency could probably be achieved by a grit trap/oil separator combination, albeit at higher cost.

5.5.11 Wetlands

As indicated previously in Section 4.3.18, the design basis and treatment capability of wetland systems for highway runoff are not adequately established at present. What little experience has been gained so far suggests that a moderate-to-high degree of water quality improvement can be achieved particularly for suspended solids and metals, and an even greater treatment is possible by combining wetland systems with other treatment or flow control stages (e.g. Meyer, 1985). Preliminary advice on design aspects can be obtained from Ellis (1991b) and other references contained therein.

5.5.12 Lagoons

In the comparative M1 study previously referred to, the lagoon gave extremely good removal efficiencies for several parameters studied (>70% for suspended solids, lead, oil and PAH), although this was partly attributed to over-design. Nevertheless, this effective approach to treatment appears to be of potential value in rural or semi-rural locations in which space constraints preclude a formal wetland or storage pond development, but still permit construction of a narrow linear lagoon which can lend itself naturally to fitting within the highway land boundary. Detailed design guidelines have not yet been developed.

5.5.13 Maintenance implications

In all of the techniques, which have been identified as being useful to remove pollution from highway runoff, the predominant action is the removal and containment of sediments. These accumulations will need to be removed at some stage if the continuing benefit of the installation is to be felt. This then presents one further problem for the maintenance authority because the material arising from such operations is likely to represent, in concentrated form, the net volume of contaminants generated on that section of road during the period since the last maintenance visit. In the case of filter drains this could amount to as much as ten years.

The material must be disposed of with adequate care. In probability it will be classified as a controlled material by the waste disposal authority and will therefore have to be transported to a licensed disposal facility.

6 Conclusions

6.1 POLLUTANTS IN RUNOFF

Highway runoff does degrade receiving water quality, though usually to only a small extent. However, there is insufficient appreciation of this fact amongst highway engineers.

Discharges from highway drainage are intermittent and can be either:

- routine discharges due to rainfall washing off pollutants that have built up over a long period from everyday use of the road
- accidental discharges.

This report mainly deals with pollution from routine discharges. More than 30 different substances have been identified in these discharges. Details are given in Section 2.

There are two main sources of pollution in routine discharges:

- pollutants from the regular passage of traffic
- pollutants from winter maintenance activities.

The major pollutants in routine discharges are of three types:

1. Sediments. These can in themselves cause an impact on the receiving water, but also act as a transport mechanism for many pollutants. Up to 85% of pollutants are to be found as, or adsorbed on, or absorbed by, sedimentary particles

2. Hydrocarbons. These are generally immiscible with water, e.g. oil

3. Metals. Although highway runoff does contain cadmium and lead, which are very toxic metals, this is at small and reducing levels. A greater problem is from the more common metals of copper, iron and zinc that are present in larger quantities.

Pollution from routine discharges, although generally of a low level, has been observed to have the following effects on receiving waters:

- increased turbidity and blanketing of stream-beds
- depletion of dissolved oxygen
- algal growth
- toxicity to stream flora and fauna
- smell, tainting and visual effects.

The potential impact of discharges on receiving waters will vary, but groundwaters are particularly vulnerable because of the great difficulty in cleaning up pollution once it is in the ground. Little is known of the fate of even common pollutants when they enter aquifers, nor is it known what processes operate inside, or in the vicinity of soakaways.

Special measures need to be taken to avoid problems from discharges to static surface waters. In particular sediment should be removed from the discharge so as to avoid the subsequent need to dredge polluted sediment.

Accidental discharges are an environmental threat because of the almost infinite number of materials that might be involved. The only satisfactory solution to this problem is containment of the spill to prevent it reaching the receiving water. This must use a method that is simple, robust and quick to operate and accessible to emergency services and disposal vehicles.

6.2 LEGISLATION

Highway authorities have a legal right to discharge to any inland waters and the pollution control authorities have only limited scope for action where highway discharges are concerned. Only if it is proven that pollution has been caused by a highway discharge can they take prescriptive action. As yet it remains unclear whether they would use such powers. This legal position is generally well understood by both parties and their agents.

The greatest current concern is the liability for the pollution of water at civil law. The Cambridge Water Company case has been decided on appeal to the House of Lords. In this case the discharger was not found liable for the pollution but only because the damage could not have been foreseen at the time of the discharge.

The problem of accidental discharge is addressed by legislation to control the construction and use of transport vehicles, and the construction of containers for dangerous substances. However, it may be possible to reduce the problem by improvements to this legislation.

Standards are now published, although not yet enacted, for the required quality of receiving waters for various uses, including fisheries and drinking water abstraction. These give a rational basis for assessing the impact of discharges.

6.3 DESIGN ISSUES

Designers have so far been more concerned with considering the quantity of discharge that may cause flooding, than with the effects of highway discharges on water quality.

There is no standard pollution assessment procedure to identify when highway drainage is likely to be a problem or what measures are needed to remedy these problems.

Research into suitable assessment and design techniques is required together with the acquisition of appropriate data. The need for future research is discussed in the following section.

Many existing drainage techniques do have an effect on pollutant discharges, and the use of best practice can avoid many problems. However, there is insufficient quantitative information on their effectiveness.

Some drainage techniques that are effective in pollution terms are not currently recommended by DOT because of operational or construction difficulties. Improved design and practice could overcome these problems. Filter drains, for example, have great potential for control of pollution, but are not currently recommended by DOT because of operational problems.

Road junctions and motorway interchanges appear to be more prone to accidental discharges than the open road. In the past no special provision for containment was made at these locations although the NRA may now request this.

6.4 MAINTENANCE

The DOT Maintenance Manual sets out schedules for cleansing and refurbishment of trunk roads and motorways and is administered by highways agents. However, there is variation between agents in the timing and scheduling of maintenance work and the only item that appears to receive regular attention is gully pot emptying. Few agents reported regular maintenance of other items or long-term programmes for major maintenance such as replacement of filter drain media.

There can be significant levels of herbicide in highway drainage discharges. The use of these substances is reducing, but is still significant.

Although highway drainage discharges can contain significant concentrations of chemicals derived from road salt, there have been very few instances where this can be shown to have caused a pollution problem in the receiving waters. The continuing improvement in methods to control the use of de-icing materials has led to a significant drop in the amount of potentially polluting material that is spread on the roads.

7 Recommendations

7.1 PROCEDURAL ISSUES

1. Highway authorities should give equal status to runoff quality considerations as to quantity requirements by adopting the design guidelines set out in Section 5 of this report.

2. Water pollution control authorities should also adopt the design guidelines set out in Section 5 of this report as a basis for assessing the suitability of highway drainage schemes.

3. Highway authorities should analyse the statistics on the risk of accidents at certain highway locations to provide guidance on places where the drainage system needs to make special provision for the containment of accidental spillages.

4. The Hazchem scheme should be kept under review to ensure that the recommended response to a spillage takes into account the environmental impact of these actions. Substances that have an environmental impact but that are not hazardous to human safety should be covered under this or a similar labelling scheme.

5. The regulations governing the transport of potentially polluting materials by road should be kept under review. This includes the regulations covering containers for materials and the construction and use of vehicles.

6. Police and fire brigade personnel should be given training in avoiding environmental impact after spillages. They should also be made aware of the assistance that the pollution regulators can give to them in this.

7.2 DRAINAGE TECHNIQUES

Reducing the pollutant load discharged from highway drains requires development work to improve current drainage techniques and to make more use of novel techniques.

1. The design and maintenance of filter drains should be developed to overcome the limitations of these systems. A first step would be to standardise on trench width to allow for the introduction of standard width machines to lift and recycle the filter material.

2. The use of cheaper local materials as filter material in filter drains should be investigated.

3. Designs should be developed for improved over-the-edge drainage systems, which make use of edge of pavement steps and of cellular geotextiles (see Section 4.3.7). The use of these designs should be encouraged in appropriate situations.

4. Comparative costings of alternative amelioration techniques should be developed to a common base to assist designers in selecting the best value system.

5. More information should be collected on the efficiencies of various drainage systems in removing pollution under critical storm conditions.

6. The use of small, less costly, oil separators, which are now available, should be encouraged by identifying the best locations for siting these in highway drainage systems and by developing standard design details.

7. A study should be carried out of the use of roadside ditches as a simple form of swale, using similar techniques to those used on full size installations.

8. The ability of sedimentation tanks to remove fine particle sizes (and hence more of the pollutants) should be studied to determine whether their use should be encouraged.

9. Constructed wetlands seem to offer a method of dealing with pollution from highway drainage. Further work should be carried out to assess their effectiveness, ideally by field trials.

10. Correctly maintained gully pots seem to be effective in reducing some pollutants, but have also been reported as increasing others. More work is needed to resolve this conflict and to recommend the maintenance regime that would be necessary for them to have a beneficial effect.

11. Further information should be collected on the rates of accumulation of contaminants in different types of drainage system. Some anecdotal information has been obtained from data extracted during maintenance work on combined filter drains, but this is random and not suitably categorised.

12. The behaviour of contaminants in and around soakaways should be studied, as existing soakaway systems will remain in use for some time.

13. The material deposited in highway drainage systems should be studied to establish its potential for continuing pollution and the hazards associated with its eventual disposal.

14. There should be research into the concentrations of MTBE and other components of unleaded fuel in highway runoff and into their passage and fate in the drainage system.

7.3 MAINTENANCE

1. Thorough and regular maintenance schedules should be set down and adhered to for all highway drainage systems. Without regular maintenance the drainage cannot operate efficiently either hydraulically or for pollutant control.

2. To ensure that the maintenance schedules are adhered to, there should be a requirement for written reports of maintenance and inspection of all highway drainage systems.

3. Future budgetary planning by highway authorities should seek to improve the standard of maintenance rather than reduce it.

4. The use of the latest frost and ice warning technology should be extended to cover all highway authorities so as to reduce unecessary use of salt and de-icing compounds.

5. The potential polluting effects of alternative de-icing materials should be investigated before they come into widespread use.

6. Development work should continue to reduce the use of herbicides to the minimum and, where appropriate, to reintroduce mechanical means for control. Where herbicides are to be used then this should be restricted to those that do not leave a harmful residue.

7. There should be further research into the passage and fate of chemicals such as herbicides and pesticides. Such research should distinguish between applications to hard surfaces and to absorbent surfaces such as vegetation and verges.

8. Strong support should be given to environmental awareness campaigns such as anti-litter, anti-fouling and re-cycling.

7.4 DESIGN AND ASSESSMENT METHODS

Guidelines for the assessment of pollutant impact from highway drainage have been set out in Section 5 of this report.

1. The application and use of the guidelines set out in Section 5 of this report should be tested by trial use on a series of typical cases. The guidelines should be modified if the testing shows this to be necessary.

2. Design runoff coefficients appropriate for the calculation of pollutant concentrations should be developed. They are likely to be different from those used for quantity considerations.

3. The average pollutant loadings given in this report should be developed to cover a greater range of AADT and to consider whether the road type or width are significant.

4. The influence of the shape of the storm on washoff of pollutants and quality of runoff should be established, together with the influence of time of concentration.

5. The sensitivity of the average pollutant loading, in the drainage system, to assumptions made as to storm intensity and antecedent dry days should be investigated.

6. Further work should be carried out to determine the quantities of oil which are present in routine runoff from highways, and the quantities which may be permitted in discharges to rivers.

References

AMY, G. et al (1974)
Water quality management, planning for urban runoff: a manual
US Environmental Protection Agency
Report No. 440/9-75-004

ARONSON, T.L. et al (1983)
Evaluation of catchbasin performance for urban stormwater pollution control
US Environmental Protection Agency
Report No. 600/2-8b-043

ASSOCIATION OF METROPOLITAN AUTHORITIES (1991)
Highway maintenance : a code of good practice : winter maintenance supplement
Association of Metropolitan Authorities (London)

AUDIT COMMISSION (1988)
Improving highway maintenance a management handbook
HMSO (London)

BARTLETT, R.E. (1976)
Surface water sewerage
John Wiley (New York)

BARTLETT, R.E. (1979)
Public health engineering. Sewerage. 2nd edition.
Applied Science Publishers Ltd. (London)

BASCOMBE, A.D, ELLIS, J.B, REVITT, D.M. and SHUTES, R.E.B. (1988)
The role of invertebrate biomonitoring for water quality management within urban catchments
Hooghart, J.H.C. (edit): *Hydrological processes and water management in urban areas*
IHP/UNBSOD, The Netherlands
404-412

BEALE, D.C. (1992)
Recent developments in the control of urban runoff.
Journal of the Institution of Water and Environmental Management. Vol.6 (No.2)
141-150

BELLINGER, E.G. et al (1982)
The character and disposal of motorway runoff water.
Water Pollution Control. Vol.8 (No.3)
372-389

BICKMORE, C. and DUTTON, S. (1984)
Water pollution and motorway runoff.
Surveyor 3 May
12-13

BICKMORE, C. and DUTTON, S. (1984)
Environmental effects of motorway runoff.
Surveyor 10 May
24-25

BICKMORE, C. and DUTTON, S. (1984)
Oil and toxic spillages in motorway runoff.
Surveyor 24 May
13-15

BORCH-JENSEN, J.E. (1978)
Runoff from motorways passing open country or rural areas.
In *Proceedings of Symposium on Road Drainage.*
OECD Rome

BOXALL A.B.A, FORROW D.M, MALTBY L.L, CALLOW P. and BETTEN C. (1993)
Toxicity of road runoff contaminated sediment to *Gammarus Pulex.*
Poster presented at *SETAC World Congress March 1993, Lisbon, Portugal*

BRITISH STANDARDS INSTITUTION (1989)
Specification for spreaders for winter maintenance.
BS 1622

BRITISH STANDARDS INSTITUTION (1983)
Code of practice for drainage of roofs and paved areas.
BS 6367

BRITISH STANDARDS INSTITUTION (1985)
Code of practice for building drainage.
BS 8301

BRITISH STANDARDS INSTITUTION (1991)
Specification for salt for spreading on highways for winter maintenance.
BS 3247

BROWN, J. R. (1973)
Pervious bitumen macadam surfacing laid to reduce splash and spray at Stonebridge, Warwickshire.
TRRL Laboratory Report 563
Transport and Road Research Laboratory

BUILDING RESEARCH ESTABLISHMENT (1991)
Soakaway design.
BRE Digest 365
Building Research Establishment (Watford)

BUTLER, D. and CLARK, P. (1993)
Sediment management in urban drainage catchments.
CIRIA Funders Report CP/11
CIRIA (London)

CATHELAIN, H. et al (1981)
Les eaux de ruissellement de chaussées autoroutières.
Bull Liason de Labs. des Ponts et Chaussées No. 116
9-24

CEDERGREEN, H.R. (1974)
Drainage of highways and airfield pavements.
John Wiley (New York)

CHILTON P.J, LAWRENCE A.R. and BARKER J.A. (1990) Chlorinated solvents in Chalk Aquifer: some preliminary observations on behaviour and transport.
In *Chalk,* Thomas Telford, London.
605-610

CIRIA (1992)
Scope for control of urban runoff
R123/124
CIRIA (London)

COLLINS, P.G. and RIDGEWAY, J.W. (1980)
Urban storm runoff quality in south east Michigan.
*Journal of the Environmental Engineering Division
ASCE Vol.106* No.EE1
153-162

COLWILL, D.M. et al (1985)
Water quality of motorway runoff.
*Transport and Road Research Laboratory
Supplementary Report 823*
TRRL (Crowthorne, Berks)

CONTROL OF PESTICIDES REGULATIONS
(1986)
HMSO (London)

CONTROL OF SUBSTANCES HAZARDOUS TO
HEALTH REGULATIONS (1988)
HMSO (London)

COOPER, P.F. and FINDLATER R.C. (Eds.) (1990)
Constructed wetlands in water pollution control.
Programme Press (Oxford)

CORBET, S.P. (1990)
Comparative trials of fin drains.
TRRL Contractor Report 221
Transport and Road Research Laboratory

CRABTREE, R.W. et al (1991)
Review of research into sediments in sewers and
ancillary structures.
FWR Report No. FRO205
Foundation for Water Research, Swindon

DAUBER L. et al (1978)
Schmutzstoffe im regenwasser kanal einer autobahn.
Stuttgarter Benchte zar Siedlungwasser wikt schaff 64
41-57

DAY, G.E. et al (1981)
Runoff and pollution abatement characteristics of
concrete grid pavements.
Bulletin 135
Virginia Water Resources Research Centre
(Blacksburg, USA)

DEGROOT, W. (1982)
Stormwater detention facilities.
American Society of Civil Engineers (New York)

DINIZ, E.V. (1976)
Quantifying the effects of porous pavements on urban
runoff. In: *Proceedings of the National Symposium of
Urban Hydrology: Hydraulics and Sediment Control*
(Lexington USA)
633-670

DOBSON, M.C. (1991)
Tolerance of trees and shrubs to de-icing salt.
Arboriculture Research Note 99/91/PATH
Department of the Environment, Arboricultural
Advisory and Information Service (Washington DC)

DoE (1988)
Circular 15/88 Environmental Assessment
HMSO (London)

DoE (1988)
Highways Considerations in Development Control
DoE Policy Guidance Note 13

DoE (1989)
*Environmental Assessment - A guide to the
procedures*
HMSO (London)

DoE (1989a)
Water Supply (Water Quality) Regulations 1989
SI 1989/1147
DoE & Welsh Office

DoE (1989b)
Surface Waters (Classification) Regulations 1989
SI 1989/1148
DoE & Welsh Office

DoE (1992)
Development Plans and Regional Planning Guidance
DoE Planning Policy Guidance Note 12

DoE (1993)
*River quality. Draft regulations: The Surface Waters
(Fisheries Ecosystem) (Classification) Regulations
1993.*
DoE & Welsh Office

DoE (1993b)
Verbal communication.

DoE/TRRL (1976)
*Road Note 35 - a guide for engineers to the design of
storm sewer systems*
HMSO (London)

DOT (1985)
*Code of practice for the routine maintenance of
highways.*
Dept. of Transport (London)

DOT (1989)
HA37/88 Hydraulic design of road edge surface water
channels.
In: *Design manual of roads and bridges Volume 4*
Section 2
Dept. of Transport (London)

DOT (1989)
HA40/89 Determination of pipe and bedding
confinement for drainage works.
In: *Design manual for roads and bridges Volume 4*
Section 2.
Dept. of Transport (London)

DOT (1991)
HA39/89 Edge of pavement details.
In: *Design manual for roads and bridges Volume 4*
Section 2
Dept. of Transport (London)

DOT (1991)
HA37/88 Hydraulic design of road edge surface water
channels. Amendment No. 1
In: *Design manual for roads and bridges Volume 4*
Section 2
Dept. of Transport (London)

DOT (1991)
*Design manual for roads and bridges Volume 3.
Highway construction details.*
Dept. of Transport (London)

DOT (1991)
Design manual for roads and bridges. Volumes 1 to 11
Dept. of Transport (London)

DOT (1992)
Trunk road maintenance manual Volume 2, Part 3 Routine and winter maintenance code.
Dept. of Transport (London)

DUSSART, G.B.J. (1984)
Effects of motorway runoff on the ecology of stream algae.
Journal of the Institute of Water Pollution Control (No.83)
409-415

EASTWOOD P.R, LERNER D.N, BISHOP P.K, and BURSTON M.W. (1991)
Identifying land contaminated by chlorinated hydrocarbon solvents.
Journal of IWEM Vol.5 No.2
163-171

ELLIS, J.B. (1979)
The natures and sources of urban sediments and their relation to water quality : A case study from N W London.
In: *Man's impact on the hydrological cycle in the UK* edited by G. Halls. Gro Abstracts Ltd. (Cambridge)

ELLIS, J.B. (1992a)
Quality issues of source control.
In: *Proceedings of CONFLO92 - Integrated catchment planning and source control*
Oxford

ELLIS, J.B. (1992)
Design criteria for managing detention basin quality.
In: *Urban Stormwater Management* edited by G. O'Loughlin. Australian Institute of Engineers (Sydney) (in Press)

ELLIS, J.B. et al (1986)
Hydrological controls of pollution removal from highway surfaces.
Water Research Vol.20 No.5
589-595

ELLIS, J.B. (1986)
Pollutional aspects or urban runoff.
In: *Urban Runoff Pollution* ed H.C. Torno et al
1-38
Springer-Verlag (Berlin)

ELLIS, J.B. (1990a)
Bioengineering design for water quality improvement of urban runoff.
In: *Developments in storm drainage* edited by D. J. Balmforth, Sheffield City Polytechnic

ELLIS, J.B. (1991a)
The design and operation of vegetation systems for urban runoff quality control.
In: *Proceedings of 3rd standing conference on Stormwater Science Control*
Coventry Polytechnic

ELLIS, J.B. (1991b)
Drainage from roads : control and treatment of highway runoff.
Report NRA 43804/MID.012
National Rivers Authority (Reading)

ELLIS, J.B, REVITT D.M, HARROP D.O, and BECKWITH P.R.(1987)
The contribution of highway surfaces to urban stormwater sediments and metal loadings.
The Science of the Total Environment (No. 59)
339-349

EXTENCE, C.A. (1978)
The effects of motorway construction on an urban stream.
Environmental Pollution No. 17
245-252

FEDERAL HIGHWAYS ADMINISTRATION (1981)
Constituents of highway runoff.
Report Nos. 8142, 8143, 8144, 8145.
FHA (Washington DC)

FEWKES, A. and JAY, A. (1992)
Use of rainwater for WC flushing.
Architect and Surveyor, Feb.
12-15

FIELD, R. (1985)
Urban runoff : pollution sources, control and treatment.
Water Res Bull Vol. 21 (No. 2)
197-206

FLETCHER I.J, PRATT C.J, & ELLIOTT G.E.P. (1978)
An assessment of importance of roadside gully pots in determining the quality of storm runoff.
In: *Urban Storm Damage* edited by P.R. Helliwell
586-602
Pentech Press (London)

FORROW D.M, BOXALL A.B.A, MALTBY L.L, CALLOW P, & BETTEN C. (1993)
The impact of road runoff on the structure and function of macro-invertebrate and microbial communities.
Poster presented at *SETAC World Congress March 1993, Lisbon, Portugal*

FOSTER et al (1991)
Mechanisms of groundwater pollution by pesticides.
Journal of Institute of Water and Environmental Management Vol.5 No.2
186-193

FROST, H. (1910)
The art of roadmaking.
The Scientific Press (New York)

FUJITA, S. (1984)
Experimental sewer system for reduction of urban storm runoff.
In: *Proceedings of the Third International Conference on Urban Storm Damage*, Goteburg, Sweden.
1211-1220

FWR (1994)
Urban pollution management manual
FR/CC002

GOFORTH, G.F. et al (1984)
Stormwater hydrological characteristics of porous and conventional paving systems.
US Environmental Protection
Agency Report No. PB84-123-728

GOMME J.W, SHURVELL S, HENNINGS S.M, and CLARK L. (1991).
Hydrology of pesticides in a Chalk catchment: surface waters.
Journal of Institution of Water and Environmental Management Vol.5 No.5
546-552.

GROTTKER, M. (1990a)
Pollutant removed by gully pots in different catchment areas.
The Science of the Total Environment (No. 93)
515-522

GROTTKER, M. (1990b)
Pollutant removal by catchbasins in West Germany. State of the art new design.
In: *Proceedings of the ASCE Engineering Foundation Conference on Urban Stormwater Quality Enhancement, Source Control, Retrofitting and Combined Sewer Technology*
Davos Platz (Switzerland)

HALL, M.J. and ELLIS, J.B (1985)
Water quality problems in urban areas.
Geo Journal Vol 11(No. 3)
265-275

HALL, M.J. et al (1993)
Design of flood storage reservoirs.
CIRIA and Butterworth-Heinemann (London)

HAMILTON, R.S. and HARRISON, R.M. (Edits) (1991)
Highway Pollution
Elseiver Science Publications (London)

HARRISON, R.M. and WILSON S.J. (1985)
The chemical composition of highway drainage waters.
The Science of the Total Environment (No. 43)
89-102

HEALTH AND SAFETY COMMISSION (1991)
Advisory Committee Report on Dangerous Substances
HMSO (London)

HEDLEY G. and LOCKLEY, J.C. (1975)
Quality of water discharged from an urban motorway.
Journal of Water Pollution Control (No. 74)
659-674

HEWITT, C.N. and RASHED, M.B. (1992)
Removal rates of selected pollutants in the runoff waters from a major rural highway.
Water Research Vol 26 (No. 3)
311-319

HILLARY, H. (1992)
Effectiveness of Highway Edgedrains
Final Report of Project No. 12, Concrete Pavement Drainage Rehabilitation, Federal Highway Administration, Offices of R&D, Washington D.C.

HIGHWAYS ACT (1854)
HMSO (London)

HOGLAND, W. (1990)
Previous asphalt construction. An overview of the situation in Sweden and the United States
In: *Infiltration and storage of Stormwater in New Developments* edited by D. Balmforth
122-135
Sheffield City Polytechnic

HOWARD and BECK (1993)
Hydro-geochemical implications of groundwater contamination by road de-icing chemicals.
Journal of Contaminant Hydrology (No. 12)
248-268

HR WALLINGFORD (1986)
Design and analysis of urban storm drainage - The Wallingford Procedure
National Water Council (London)

HRS (1984)
The drainage capacity of BS road gullies and a procedure for estimating their spacing
TRRL/Hydraulics Research Station Contractors Report 2

HVITVED-JACOBSEN, T. et al (1986)
Fate of phosphorus and nitrogen in ponds receiving highway runoff.
The Science of the Total Environment (No. 33)
259-270

HYDRAULICS RESEARCH STATION (1989)
The drainage capacity of BS road gullies and a procedure for estimating their spacing.
TRRL Contractors Report 2
TRRL

KARUNARATNE, S.H.P.G. (1992)
The influence of gully pot performance on the entry of sediment into sewers.
Unpublished PhD thesis
South Bank University (London)

KNOTT, G.E. and TAYLOR, A.J. (1985)
The use of detention tanks for attenuation - a review of current practice.
Report No. 171E
Water Research Centre (Swindon)

LAGER, J.A. et al (1977)
Catchbasin technology overview and assessment.
US Environmental Protection Agency Report No. FPA 600/2-77-051

LANGE, G. (1990)
The design and construction of treatment processes for highway runoff in the FRG.
The Science of the Total Environment (No. 93)
499-506

LAWRENCE A.R. and FOSTER S.S.D. (1991)
The legacy of aquifer pollution by industrial chemicals: technical appraisal and policy implications.
Quarterly Journal of Engineering Geology, Vol.24 (No.2)
231-239

LAXEN, D.P.H. and HARRISON, R.M. (1977)
The highway as a source of water pollution.
Water Research (No. 11)
1-11

LEONARD, O.J. and SHERRIFF, J.D.F. (1992)
Scope for control of urban runoff Volume 3 : guidelines
CIRIA Report No. 124

LITTLE, P. and WIFFIN, R.D. (1977)
Emission and deposition of petrol engine exhaust Pb - 1. Deposition of exhaust Pb to plant and soil surfaces.
Atmos Environ No. 11
437-447

LITTLE, P. and WIFFIN, R.D. (1978)
Emission and deposition of lead from motor vehicle exhaust II. Airborne concentration, particle size and distribution of lead near motorways.
Atmos Environ No. 12
1331-1341

LOCAL AUTHORITY ASSOCIATIONS (1989)
Highway maintenance a code of good practice.
Association of County Councils (London)

LORD, B.N. (1987)
Non-point source pollution from highway stormwater runoff.
The Science of the Total Environment (No. 59)
437-446

MAESTRI, B. et al (1988)
Managing pollution from highway stormwater runoff.
Transportation Research Record (No. 1166)
15-21

MALMQUIST, P.A. (1978)
Atmospheric fallout and street cleaning effect on urban stormwater and snow.
Prog. Water Tech No. 10
495-505

MALMQUIST, P.A. (1983)
Urban stormwater pollutant sources.
Chalmers University of Technology (Gothenburg)

MANCE, G. (1981)
The Quality of Urban Storm Discharges
Report 192-M
Water Research Centre (Stevenage)

MANCE, G. and HARMAN, M.N.I. (1978)
The quality of urban stormwater runoff.
In: *Urban Storm Damage* edited by R.P. Helliwell
603-618
Pentech Press (London)

MARTIN, H.E. and SMOOT, J.C. (1988)
Constituent land changes in urban stormwater runoff routed through a detention pond wetland system in Central Florida.
Journal of Environmental Engineering ASCE Vol.114
226-249

MASKELL, A.D. and SHERRIFF, J.D.F. (1992)
Scope for control of urban runoff Volume 2. A review of present methods and practice.
CIRIA Report 124

MEDHURST (1992)
Verbal communication.
Three Valleys Water plc.

MEYER, J. (1985)
A detention basin/artificial wetland treatment system to renovate stormwater runoff from urban highway and industrial areas.
Wetlands No. 5
135-146

MORRISON, G.M.P. et al (1988)
Transport mechanisms and processes for metal species in a gullypot system.
Water Research Vol. 22 (No. 11)
1417-1427

MUSCHACK, W. (1990)
Pollution of street runoff by traffic and local conditions.
The Science of the Total Environment (No. 93)
419-431

NATIONAL RIVERS AUTHORITY (1992)
Policy and practice for the protection of groundwater.
National Rivers Authority (Bristol)

NIEMCZYNOWICZ, J. (1989)
Swedish way to stormwater enhancement by source control.
In: *Proceedings of ASCE Engineering Foundation Conference on Urban Stormwater Quality Enhancement, Source Control, Retrofitting and Combined Sewer Technology*
Davos Platz (Switzerland)

NOBLE and COOK (1987)
Motorway de-icing - Urea the panacea.
Journal of the Institution of Water Pollution Control

NOVOTONY, V. (1984)
Efficiency of low cost practices for controlling pollution by urban runoff.
3rd International Conference on Urban Storm Drainage, Goteberg, Sweden

NOVOTNY, V. and CHESTERS, G. (1981)
Handbook of non-point pollution: sources and management.
Van Nostrand Reinhold (New York)

OSBORNE, M.P. and PAYNE, J.A. (1992)
Pollution control
In *Drainage design*, edited by P. Smart and J.G. Hebertson, Blackie (Glasgow)
151-168

PAYNE, J.A. and WATKINS, D.C. (1992)
The CIRIA infiltration manual.
In: *Proceedings of CONFLO92 Integrated catchment planning and source control* (Oxford)

PEARSON, D. (1990)
Highway drainage and its problems.
In: *Developments in Storm Drainage : A symposium on Infiltration and Storage of Stormwater in New Developments* edited by D. Balmforth
Sheffield City Polytechnic

PISANO, W.C. et al (1979)
Dry weather deposition and flushing for CSO pollution control.
US Environmental Protection Agency
Report No. 600/2-79-133

PITT, R. (1979a)
Best management practice; implementation
US Environmental Protection Agency Report No. EPA 905/9-81-002

PITT, R. (1979b)
Demonstration of non-point pollution abatement through improved street cleaning practice.
US Environmental Protection Agency Report No. 600/2-79/161

PITT, R. (1985)
Characterising and controlling urban runoff through street and sewerage cleaning.
US Environmental Protection Agency Report No. 600/S2-88/038

PRATT, C.J. (1989)
Permeable pavements for stormwater quality enhancement.
In: *Proceedings of ASCE Engineering Foundation Conference on Urban Stormwater Quality Enhancement, Source Control, Retrofitting and Combined Sewer Technology.*
Davos Platz (Switzerland)

PRATT, C.J. (1992)
Quality issues of source control.
In: *Conflo 92. Integrated Catchment Planning and source control* (Oxford)

PRATT, C.J. et al (1986)
Role of highway gullies in determining water quality in separate storm sewers.
In: *Proceedings of the International Conference on Urban Storm Water Quality and Effects on Receiving Waters*
Wageningen, Netherlands

PRATT, C.J. et al (1987)
Suspended solids discharge from highway gully pots in a residential catchment.
The Science of the Total Environment (No. 59)
355-364

PRATT, C.J. et al (1989)
Urban stormwater reduction and quality improvement through the use of permeable pavements.
Water Science Technology Vol. 21
769-778

PRATT, C.J. and ADAMS, R.J.W. (1984)
Sediment supply and transmission via roadside gully pots.
Science of the Total Environment (No. 33)
213-224

PRATT, C.J. and HOGLAND, W. (1990)
Permeable pavements : design and maintenance.
In: *Developments in Storm Drainage : A symposium on Infiltration and Storage of Stormwater in New Developments* edited by D. Balmforth
Sheffield City Polytechnic

PRAVOSHINSKY, N.A. (1975)
Basic principles for determining regulating structure parameters to prevent rain sewer receivers from contamination.
Progress on Water Technology Vol 7 (No. 2)
301-307

PRICE M, ATKINSON T.C, WHEELER D, BARKER J.A. and MONKHOUSE R.A. (1992)
A tracer study of the threat posed to a chalk aquifer by contaminated highway runoff.
Institution of Civil Engineers: Water, Maritime and Energy Journal 1 (1)
In press.

PRICE, M. et al (1989)
Highway drainage to the chalk aquifer.
British Geological Survey Report No. WD/89/3
British Geological Survey (Nottingham)

RANCHER, J. and RUPERD, M. (1983)
Moyen d'action pour limiter la pollution d'eaux de ruissellement en systeme separatif et unitaire.
Tribune Cebedeau Vol.36. No.472
117-130

REVITT M.O, LERNER D.N, and LLOYD J.W. (1990)
Chlorinated solvents in UK aquifers.
Journal of Institution of Water and Environmental Management Vol.4 No.3
242-250.

RUPERD, M. (1987)
Efficacité des ouvrages de traitement des eaux de ruissellement.
Science Technique l'Urbanisme
Div. Equip Urbains (Paris)

SALEEM, Z.A. (1977)
Road salts and quantity of groundwater from a dolomite aquifer in the Chicago area.
In: *Hydrologic problems in Kast regions* edited by Dilamarter and Callay
West Kentucky Uni Press
364-369

SARTOR, J.D. and BOYD, G.B. (1972)
Water pollution aspects of street surface contaminants.
US Environmental Protection Agency Report EPA R2/72-081 (Washington DC)

SAUL, A.J. and ELLIS. D.R. (1992)
Sediment deposition in tanks.
Water Science and Technology Vol.25(No. 8)
189-198

SCOTTISH OFFICE ENVIRONMENT DEPARTMENT (1992)
Draft Code of Practice on Conservation, Access and Recreation for Water, Sewerage and River Purification Authorities

SEWERS AND WATER MAINS COMMITTEE (1991)
A guide to sewerage operational practices.
Water Services Association/Foundation for Water Research

SPALDING, R. (1970)
Selection of materials for sub-surface drains.
TRRL Laboratory Report 346
TRRL (Crowthorne)

STOTZ, G. (1987)
Investigations of the properties of surface water runoff from Federal Highways in the FRG.
Science of the Total Environment (No. 59)
329-337

STOTZ, G. (1990)
Decontamination of highway surface runoff in the FRG.
Science of the Total Environment (No. 93)
507-514

STRECKER, E.W. et al (1990)
The US Federal Highway Administration receiving water impact methodology.
Science of the Total Environment (No. 93)
489-498

SUDA, S. et al (1988)
Development and application of permeable paving concrete block.
In: *Proceedings of the Third Conference on Concrete Block Paving, Rome*
130-137

SZATKOWSKI and BROWN, J.R. (1977)
Design and performance of previous wearing courses for roads in Britain 1967-1976.
Highways and Road Construction International Jan-Feb
12-16

TESTER D.J. and HARKER R.J. (1981)
Groundwater pollution investigations in the Great Ouse basin.
Journal of IWPC Water Pollution Control, Vol.80
614-631

TODD D.K. 1980
Groundwater Hydrology 2nd Edition,
John Wiley, New York.

TRANSPORT AND ROAD RESEARCH LABORATORY (1969)
Hydraulic efficiency and spacing of BS road gullies.
TRRL Laboratory Report 277
TRRL (Crowthorne)

TRANSPORT AND ROAD RESEARCH LABORATORY (1973)
Drainage of level or nearly level roads.
TRRL Laboratory Report 602
TRRL (Crowthorne)

TRANSPORT AND ROAD RESEARCH LABORATORY (1984)
The drainage capacity of BS road gullies and a procedure for estimating their spacing.
TRRL Contractor Report 2
TRRL (Crowthorne)

TRANSPORT AND ROAD RESEARCH LABORATORY (1987)
Contractors Report 76 - Methods for the removal of surface water from rural trunk roads
TRRL (Crowthorne)

VAN DE VEN, F.H.M, NELEN, A.J.M, and GELDOF, G.D.(1992)
Urban Drainage.
In *Drainage design* edited by P.Smart and J.G. Herbertson. Blackie, Glasgow and London.
118-150

VERNIERS, G. and LOZE, H. (1985)
Etude écologique des bassins d'orage autoroutiers.
In: *Ann. Travaux Publics de Belgique, No. 2*
14-21

WRc (1993)
Surface water outfalls, quality and environmental impact management.
Report No. VM 1400
WRc (Swindon)

WALLER, D.H. and HART, W.C. (1986)
Solids, Nutrients and Chlorides in Urban Runoff.
In: *Urban Runoff Pollution* (ed.) Torno et al NATO ASI Series. Vol. G10, Springer-Verlag

WATKINS, D.C. (1991)
The hydraulic design and performance of soakaways.
H R Wallingford Report No. SR 271

WATKINS, L.H. and FIDDES, D. (1984)
Highway and urban hydrology in the tropics.
Pentech Press (London)

WELHAM, K.P and HUNT, G.(1992)
Highway Drainage - Design Guide
Oxfordshire County Council

WINTERS, G.R. and GIDLEY, J.L. (1980)
Effects of Roadway Runoff on Algae
Report RHNA/CA/TL - 80/24
Office of Transport Laboratories, Department of Transportation, Federal Highway Administration (Washington DC)

YOUSEF, Y.A. et al (1985)
Removal of highway contaminants by roadside swales.
Trans Res REc (No. 1017)
62-68

YOUSEF, Y.A, HUITVED-JACOBSEN, T, VANIELISTA, M.P. and HARPER, H.H. (1987)
Removal of contaminants in highway runoff through swales
Science of the Total Environment (Vol 59)
391 - 399

Appendix A Development of numerical method for surface waters

A.1 BACKGROUND TO THE METHOD

A.1.1 Scope

The numerical method given in the guidelines in this report consists of a series of simple calculations that give an approximate comparison between the water quality standards for the receiving water and the concentration of pollutants resulting from the highway discharge.

This method is intended for assessment of the pollutant impact of metals in the highway runoff (particularly copper and zinc) although it could be adapted for other pollutants that do not decay in the receiving water.

The method is not suitable for consideration of the impact of oxygen demanding pollutants or ammonia. For these pollutants highway drainage is likely to add to problems caused by other, larger sources of the pollutants and all sources will have to be considered together. Also for these pollutants it is necessary to consider their rate of decay in the receiving water and short-term problems that may occur during this process.

A.1.2 Outline of the method

The receiving water standards for the pollutants considered by the model are set as 95% values, that is, the concentration that is only to be exceeded for 5% of the time. A full analysis of this would require a model of the build-up, washoff and dilution of pollutants and the use of such a model to analyse a long record of rainfall and river flows to determine the 95% value. This type of analysis is not worthwhile except in very exceptional circumstances. A simpler method of achieving the same aims is required. Such a method can be based on the following assumptions:

1. That problems of impact on the receiving water are most likely to occur in summer. At this time the river flows will be low, so giving less dilution and there will have been a long period of dry weather for pollutants to build up on the road surface. Therefore it can be assumed that the river will be at its low flow condition, and that this flow will not increase significantly due to runoff from the natural catchment during a storm that could wash pollutants from the road surface.

2. That careful choice of a single rainstorm with an appropriate pollutant build-up time will allow this storm to give the same results as the 95% concentration from the full series.

It is likely that the choice of storm will depend on some or all of the following factors:

- the region of the country
- the dilution in the river
- the pollutant being considered
- the assumptions about pollutant washoff behaviour
- the time of concentration of the drainage system
- the duration of storm.

The method was developed by comparing the results for modelling long data series with modelling individual rainstorms so as to identify which of these factors did have to be considered, and how they were related to the choice of storm.

A.2 DEVELOPMENT OF THE METHOD

A.2.1 Pollutant washoff model

Butler and Clarke (1993) put forward a model for the build-up and removal of pollution on the catchment surface.

$$\frac{dM}{dt} = a - bM$$

1

where   M   surface load (g/m²)
        t   time
        a   rate of supply
        b   removal constant

The removal is due to a variety of effects, but the principal one is usually rainfall. It is sensible, and in fact has been demonstrated, that this removal is a function of the rainfall intensity. Various authors have proposed forms for this function, but a common one is that the removal is proportional to rainfall intensity. (Other forms have the rainfall intensity raised to a power of 1.5 or 2.0). For the purposes of this study a power of 1.0 was assumed.

$$\frac{dM}{dt} = a - kiM$$

2

where   i   rainfall intensity
        k   a removal constant

Butler and Clarke quote average removal constants per week, but do not relate these to rainfall. These values are therefore in effect average values for the term ki. They quote 0.05 to 0.25 / week for sites in London. The lower values are assumed by Butler and Clarke to be due to rainfall removal, with the higher values being the combination of rainfall and sweeping. (All of their sites were swept regularly.) If we assume that we can relate this to the average rainfall intensity over a week then we can derive a value for k.

Annual rainfall over London is typically 700 mm, giving an average rainfall of 13.5 mm/week. Therefore k has a typical value of 0.0033 /mm but with values of up to 0.0185 /mm if they include the effects of sweeping as well as rainfall.

Using equation 2 implemented in a simple finite difference model, we can calculate the amount of pollutant washed off (W) in each timestep of a long rainfall series as:

$$W = M * ki\delta t$$

3

Concentration of pollutant in the discharge is given by this divided by runoff volume. This study assumed a fixed 50% runoff (PR). This is lower than would be used for the design of the hydraulic capacity of a drainage system, but for pollution the critical case is the lowest dilution rather than the maximum runoff volume.

$$C = \frac{M * k}{PR}$$

4

However, equation 4 would give very odd results as they would be biased to very small or even zero flows with high concentrations. Therefore we should assume dilution with a baseflow in the river (Q_r). The resulting change in concentration in the river (C_r) can therefore be calculated.

$$\Delta C_r = \frac{M * ki}{iPR + Q_r} \qquad 5$$

The actual concentration depends on the build-up rate, which is different for different road types and pollutants. However, the storms and antecedent conditions that cause this concentration are independent of this. The analysis can therefore be carried out for a single build-up rate of 1 kg/ha/a and the results factored by actual build-up rate.

A.2.2 Test data sets

Three long rainfall records were available for different regions of England with different rainfall characteristics. These were:

- Site A - representative of the dry east of England
- Site B - representative of the wetter west of the country
- Site C - representative of a large area of central and southern England.

Each record is available as both hourly and daily rainfall depths. The length of the records is typically 20 years. A full range of analysis was carried out on the Site A data including sensitivity analysis. More limited analysis was then carried out on the other two data sets to determine the regional differences from the Site A results.

A.2.3 Initial analysis

Initial analyses were carried out using the washoff model for a road area of 1 ha and river base flows of 1 l/s and 10 l/s using a 20 year record of daily rainfall data for Site A. The river flows are equivalent to about 0.5 and 5.0 mm/hr of rainfall respectively. This analysis showed the following results:

1. The mass of pollutant on the catchment surface varied between about 3000 and 7000 hours' build-up with an average of about 4500 hours (about six months).

2. The mass of pollutant removed in any storm is small compared to the amount on the catchment surface; typically less than 10%.

3. The daily rainfall that causes the 95% values of river concentration is typically between 8 and 12mm.

4. The mass of pollutant removed in the storms that cause the 95% values of river concentration is equivalent to about 120 hours' build-up.

This analysis suggested that the best way of representing river flow is by the dilution that it provided to the runoff from these storms that were appropriate for the 95% concentrations. A suitable range of dilutions to investigate was chosen as two, six and 12 times the runoff volume of a rainfall of 10mm/day.

A.2.4 Sensitivity to timestep

Using the Site A rainfall series and these dilution ratios the differences in results between carrying out the analysis using daily data and hourly data was checked.

Table A.1 *Pollutant concentrations for daily and hourly data for Site A (dilution 6:1)*

Timestep	90%	95%	99%
day	0.027	0.043	0.077
hour	<0.015	0.035	0.134

The analysis showed that for the most extreme values, those that are only exceeded for 1% of the time, the hourly concentration is higher than the daily concentration. This is reasonable as high concentrations are likely to last only a short period. However, for more frequent events the hourly concentration is less than the daily concentration. This is because the concentration tends to zero for no rain, and as the rain tends to come in short bursts separated by dry periods, there is a larger proportion of hours when it does not rain than of days when it does not rain. The hourly values therefore tend to zero more quickly than the daily values. At the 95% level at which the standards are set the difference between hourly and daily values is small.

Discussions with the NRA confirmed that the use of daily values would be adequate for comparison with the quality standards.

A.2.5 Sensitivity to erosion rate

A sensitivity analysis was carried out to see the effect of changing the erosion rate of the washoff equation. This should have very little effect on the results as once the equilibrium condition is reached, all of the storms in a year in total wash off all of the pollutant deposited in a year. Changes in the erosion rate simply change the pollutant mass on the catchment at which equilibrium is reached.

The comparison was made between the default value of 0.0033 mm^{-1} and 0.01 mm^{-1}. The resulting concentrations in the river changed by less than 2%, with those with larger dilution ratios showing no change at all. The river concentrations are therefore not sensitive to the assumption about erosion rate. The results also showed that although the order of the individual storms did change with the value of k, the size of storms producing the 95% values did not change.

A.2.6 Identification of critical storms for Site A

From the ranked list of results it is possible to read the size of storm and the amount of pollutant washed off to produce the 95% concentrations. The concentration does not change rapidly from storm to storm in the ranked list, and for ten or 20 results either side of the 95% point the concentrations in the river are very similar. However, these storms all have slightly different depths of rainfall and amounts of pollution washed off. The range in rainfall depth is typically 8 to 10 mm. The characteristic rainfall values were therefore taken by averaging over 50 values centred on the 95% point.

Table A.2 *Classification of critical storms for Site A*

Dilution	Daily rainfall (mm)	Build-up time for pollutant washed off (hr)
2	8.90	129.0
6	8.46	127.5
12	8.64	127.7

These figures show only very small variation with different dilution ratios, and can be taken as independent of dilution.

A.2.7 Analysis of Site B data

A restricted set of the same analyses was run for the Site B data to represent the wetter west of the country. The results show similar patterns to the Site A data but the concentrations are typically 80% of the Site A values. The storms were classified in the same way as for Site A by averaging over 50 storms centred on the 95% value.

Table A.3 *Classification of storms for Site B*

Dilution	Daily rainfall (mm)	Build-up time for pollutant washed off (hr)
2	11.75	115.0
6	11.60	114.0
12	11.40	114.0

As for Site A the storm characteristics show only very small variation with dilution ratio and can be considered to be independent of this. The storms are larger and the build-up times shorter than the Site A results, as might be expected for a wetter region.

A.2.8 Analysis of the Site C data

A limited set of analyses was carried out on the rainfall record from Site C. Only the dilution ration of 1:6 was analysed in detail. The concentrations are slightly lower than those from Site A. The storms were classified as before and show that the rainfall characteristics are slightly different from Site A with larger rainfall depth but slightly shorter build-up times.

Table A.4 *Classification of storms for Site C*

Dilution	Daily rainfall (mm)	Build-up time for pollutant washed off (hr)
6	11.30	123.0

A.3 DEVELOPMENT OF A GENERAL METHOD

The analysis showed that a standard storm with standard duration of pollutant build-up can represent the 95% value of the change in river concentration. The storm and build-up period are independent of: the build-up rate of the pollutants, the washoff rate of the pollutants and the dilution in the river (at least for the range of dilutions considered here). The sensitivity of the results to the power index of the washoff equation has not been tested.

The build-up time is very similar for all three of the regions that have been analysed and can be taken as a constant 120 hours (five days). The depth of rain does vary regionally and can be summarised as

A	8.7 mm
B	11.6 mm
C	11.3 mm

The results of the analysis can now be used to produce a general assessment method that can be used for any part of the UK without having to obtain a long series of rainfall data. The method will involve making a simple calculation of the amount of pollutant build-up in five days and the dilution provided by the river to a specified depth of rainfall in a 24 hour period.

A.3.1 Rainfall statistics

The depth of rain found from the analysis must be related to the Met Office rainfall statistics for each area so that appropriate depths of rain can be derived for other areas.

The storms that are being used are of 24 hour duration and of return period less than one year. Return periods of less than one year cannot conveniently be defined using the Met Office's statistics, and so a relationship with the 1 year 24 hour rainfall depth is the most appropriate.

These depths of rain are calculated using the graphical method given in Volume 1 of the Wallingford Procedure and the A4 maps in the same publication.

Table A.5 *Comparison with rainfall statistics*

	Site A	Site B	Site C
M5_60_min (mm)	20	19	19.5
r	0.4	0.32	0.33
z1	2.2	2.85	2.6
M5_24hr (mm)	44.0	54.1	50.7
z2	0.71	0.72	0.72
1 year 24 hour depth (mm)	31.2	39.0	36.5
Depth found in analysis (mm)	8.7	11.6	11.3
% of 1 year 24 hour depth	28	30	31

A standard relationship of 30% of the 1 year 24 hour storm is assumed. This is valid for the regions that have been analysed but may show some variation in other regions. This relationship was then used to derive a map showing the depth of rain to be used for each area of the country. This was based on the maps given in Volume 1 of the Wallingford Procedure. For each value of the rainfall ratio (r) the maps were inspected to determine the average value of the 1 in 5 year 60 minute rainfall depth (M5_60_min). These values were used to calculate the rainfall depth to be used in the method as defined above. For values of r that predominately occurred in Scotland and Northern Ireland the values of z2 appropriate to these regions were used. This map is given as Figure 5.4 in the report.

A.3.2 Testing general method against derived results

Developing this general method has involved some rounding of values in assuming a standard pollutant build-up time and in standardising on 30% of the one year storm. The method was therefore compared with the results from the analysis of the three sites to ensure that this rounding had not introduced significant errors.

For each of the three sites the method was used to calculate the depth of rain, the concentration in the river (using the actual river flows used in the main analysis) and this was compared to the results for river concentrations derived from the analyses of the full rainfall series. The results are shown in Table A.6.

The figures for Site A are slightly underestimated due to the rounding down of the build-up time. However, the errors are small compared to the other unknowns and can be accepted.

Table A.6 *Pollutant concentrations for general method and rainfall series analysis*

		Site A	Site B	Site C
Rainfall depth (mm)		9.3	11.7	11.0
River flow				
0.00116 l/s (100 m³/day)	single	0.093	0.086	0.088
	series	0.101	0.083	-
0.00347 l/s (300 m³/day)	single	0.039	0.038	0.038
	series	0.043	0.036	0.039
0.00694 l/s (600 m³/day)	single	0.021	0.021	0.021
	series	0.023	0.020	-

A.3.3 Comparison with water quality standards

The method based on a single storm has been shown to give a good representation of the river concentrations at the 95% level. It is, therefore, now possible to use this method to develop a table of guidelines that can be used to identify when there may be a problem of unacceptable concentrations of dissolved pollutants or those attached to fine sediment.

The table is simply to indicate when further investigation should be carried out. It is therefore appropriate to base it on a worst case situation to ensure that all problem cases receive additional investigation. The worst case in rainfall terms is the area around Thetford (East Anglia), which is the driest area of the country with the greatest risk of short high intensity storms. The appropriate daily rainfall for Thetford is 8 mm.

The initial analysis indicated that the critical pollutant was dissolved copper as this was the standard most likely to be breached by highway discharges. The standards for dissolved copper vary with the hardness of the water - a hardness of 50-100 mg/l $CaCO_3$ was assumed giving a standard of 40 µg/l. The method must allow for a background concentration in the river upstream of the discharge, and again a pessimistic assumption is made that this would be equal to half of this standard. The permitted increase in concentration due to the highway discharge is therefore 20 µg/l.

Table A.7 *Comparison of predicted concentrations with fisheries standards for copper*

Traffic count veh/day	Build-up rate (kg/ha/a)	Dilution					
		2	3	4	6	12	16
< 5000	0.2	0.018	0.013	0.011	0.008	0.005	0.003
5000 - 15 000	0.3	**0.027**	0.020	0.016	0.012	0.006	0.005
15 000 - 30 000	0.4	**0.036**	**0.027**	0.022	0.018	0.011	0.007
> 30 000	1.2	**0.110**	**0.082**	**0.066**	**0.047**	**0.025**	0.020

Figures shown in **bold** exceed the permitted increase, and drainage discharges that fall into this area of the table will require an assessment of the resulting concentration using the method as developed above.

Note that this table is not applicable for a hardness in the receiving water of less than 50mg/l $CaCo_3$. For these waters the method developed above must be used whatever the dilution and road size.

Further details of the work described in this Appendix are available from CIRIA in a supplementary report.

Appendix B Highway drainage assessment worksheet

	Description	Notes	Value
1	5% river flow (m3/s)	Obtain from regulator	
2	5% river flow (m3/day)	[1] * 3600 * 24	
3	FE class (1-6)	Obtain from regulator	
4	Hardness (mg/l CaCO3)	Obtain from regulator	
5	Road width (m)	Width of impermeable surface of new road	
6	Road length (km)	Length of new road to this outfall	
7	Extra for junctions (m2)	Allowance for extra impermeable area	
8	Road area (m2)	1000 * [5] * [6] + [7]	
9	Traffic flow (veh/day)	Design traffic flow	
10	Runoff coefficient	Use 0.5 unless justification for other value	
11	Rainfall depth (mm)	Read from Figure 5.2	
12	Runoff volume (m3)	[8] * [10] * [11] / 1000	
13	Dilution	[2] / [12]	
14	Impact code	From Table 5.6 for values in [2] [9] & [13]	
15	Total road area (m2)	All roads draining to this river reach with traffic flow > 5000	
16	Average traffic flow (veh/day)	For roads included in [15]	
17	Runoff volume (m3)	[15] * [10] * [11] / 1000	
18	Dilution	[2] / [17]	
19	Impact code	From Table 5.6 for values in [2] [16] & [18]	
20	Overall impact code	Worst case of [14] & [19]	
	Detailed assessment	Carry out if [20] includes "D"	
21	Annual build up of soluble copper	From Table 5.1 for value in [16]	
22	5 day build-up	5 * [21] / 365	
23	Standard for copper (mg/l)	From Table 5.5 / 1000 for values in [3] & [4]	
24	Upstream conc of copper (mg/l)	From regulator or [23] / 2	
25	Change in conc of copper (mg/l)	1000 * [22] / ([17] * (1 + [18]))	
26	Downstream conc of copper (mg/l)	[24] + [25]	
27	Is there an impact?	Yes if [26] > [23]	
28	Annual build-up total zinc (kg)	From Table 5.1 for value in [16]	
29	5 day build-up	5 * [28] / 365	
30	Standard for zinc (mg/l)	From Table 5.5 / 1000 for values in [3] & [4]	
31	Upstream conc of zinc (mg/l)	From regulator or [30] / 2	
32	Change in conc of zinc (mg/l)	1000 * [29] / ([17] * (1 + [18]))	
33	Downstream conc of zinc (mg/l)	[31] + [32]	
34	Is there an impact?	Yes if [33] > [30]	

B.1 WORKED EXAMPLE

An example of the use of the worksheet is given below to illustrate the application of the guidelines. The scheme being considered is for a new road which is to be constructed near Bedford and drained into a small river. There is already a small length of road draining into the river.

The new road is a dual two lane, all purpose road with a design traffic flow of 40 000 veh per day. The width of impermeable surface for the road is 20.5 m and the length to be drained into the river is 2450 m. This length includes a junction with the existing road. The junction occupies an additional 2510 m^2 of impermeable surface. The existing road, which also drains into the same stretch of river, is a single two lane road with an average daily traffic flow of 10 000 vehicles. The length of the existing road draining into the river is 726.5 m.

The river has a 5% low flow of 0.016 m^3/s. It is a reasonable fishing river; class 1b on the old NWC classification, FE class 2 on the new scheme. It is a lowland river with a hardness of 610 mg/l. Some information is available on concentrations of copper and zinc upstream of the road drainage discharges although this is based on very few samples. The average concentrations are:

    dissolved copper 0.040 mg/l
    total zinc   0.120 mg/l

This provides all the data which are needed to work through the guidelines, and the calculations are shown in the worksheet.

For item (16) in the worksheet the average traffic flow is calculated by multiplying the impermeable area by the traffic flow for each road and then dividing by the total impermeable area.

The calculations illustrate that the initial simple assessment shows that oil separation and detailed assessment of pollutants is needed for both the individual outfall and the combined effect of both outfalls. However, the assessment of pollutants shows that no special measures are needed for copper and zinc.

The scheme should be constructed with oil separators and consideration should be given to containment facilities for dealing with spillages. No abatement measures are needed for other pollutants.

	Description	Notes	Value
1	5% river flow (m3/s)	Obtain from regulator	0.016
2	5% river flow (m3/day)	[1] * 3600* 24	1382.4
3	FE class (1-6)	Obtain from regulator	2
4	Hardness (mg/l CaCO3)	Obtain from regulator	610
5	Road width (m)	Width of impermeable surface of new road	20.5
6	Road length (km)	Length of new road to this outfall	2.450
7	Extra for junctions (m2)	Allowance for extra impermeable area	2510
8	Road area (m2)	1000 * [5] * [6] + [7]	52 735
9	Traffic flow (veh/day)	Design traffic flow	40 000
10	Runoff co-efficient	Use 0.5 unless justification for other value	0.5
11	Rainfall depth (mm)	Read from Figure 5.2	9
12	Runoff volume (m3)	[8] * [10] * [11] / 1000	237
13	Dilution	[2] / [12]	5.83
14	Impact code	From Table 5.6 for values in [2] [9] & [13]	D O
15	Total road area (m2)	All roads draining to this river reach with traffic flow >5000	60 000
16	Average traffic flow (veh/day)	For roads included in [15]	36 000
17	Runoff volume (m3)	[15] * [10] * [11] / 1000	270
18	Dilution	[2] / [17]	5.12
19	Impact code	From Table 5.6 for values in [2] [16] & [18]	TD O
20	Overall impact code	Worst case of [14] & [19]	D O
	Detailed assessment	Carry out if [20] includes "D"	YES
21	Annual build-up of soluble copper	From Table 5.1 for value in [16]	1.2
22	5 day build-up	5 * [21] / 365	0.0164
23	Standard for copper (mg/l)	From Table 5.5 / 1000 for values in [3] & [4]	0.112
24	Upstream conc of copper (mg/l)	From regulator or [23] / 2	0.040
25	Change in conc of copper (mg/l)	1000 * [22] / ([17] * (1 + [18]))	0.0099
26	Downstream conc of copper (mg/l)	[24] + [25]	0.05
27	Is there an impact?	Yes if [26] > [23]	NO
28	Annual build-up total zinc (kg)	From Table 5.1 for value in [16]	5.0
29	5 day build-up	5 * [28] / 365	0.068
30	Standard for zinc (mg/l)	From Table 5.5 / 1000 for values in [3] & [4]	0.500
31	Upstream conc of zinc (mg/l)	From regulator or [30] / 2	0.120
32	Change in conc of zinc (mg/l)	1000 * [29] / ([17] * (1 + [18]))	0.041
33	Downstream conc of zinc (mg/l)	[31] + [32]	0.161
34	Is there an impact?	Yes if [33] > [30]	NO